ED LEWIS

MORE DAILY WORKOUTS

FOUNDATION TO YEAR 6

BUILDING VISUALISATION SKILLS
IN MEASUREMENT, GEOMETRY AND SPACE

OXFORD
UNIVERSITY PRESS

Oxford University Press is a department of the University of Oxford.
It furthers the University's objective of excellence in research,
scholarship, and education by publishing worldwide. Oxford is a registered
trademark of Oxford University Press in the UK and in certain other countries.

Published in Australia by
Oxford University Press
Level 8, 737 Bourke Street, Docklands, Victoria 3008, Australia.

 A catalogue record for this
book is available from the
National Library of Australia

ISBN 978 0 19 033855 8

Reproduction and communication for educational purposes
The Australian *Copyright Act 1968* (the Act) allows educational institutions that
are covered by remuneration arrangements with Copyright Agency to
reproduce and communicate certain material for educational purposes.
For more information, see copyright.com.au.

Editing consultancy by Clare Way
Illustrated by Marty Schneider, Martian Studio
Produced by Newgen KnowledgeWorks Pvt. Ltd.
Printed in China by Golden Cup Printing Co, Ltd.

*Links to third party websites are provided by Oxford in good faith and for information only.
Oxford disclaims any responsibility for the materials contained in any third party website
referenced in this work.*

Contents

OXFORD UNIVERSITY PRESS

Introduction

About *More Daily Workouts*

More Daily Workouts is a primary mathematics resource that supports the development of measurement and geometry skills, concepts and understandings. As a companion publication to the second edition of *Daily Workouts*, it offers a comprehensive range of daily activities that foster visual thinking and mental imagery. It is also aligned with the Australian Curriculum: Mathematics, Measurement and Space strand, incorporating the proficiencies of understanding, fluency, problem-solving and reasoning.

What is mental imagery?

Mental imagery, or spatial/visual thinking, includes the ability to form and retain mental images of objects and their location, as well as to manipulate them by imagined movement, rotation or transformation. Thinking in images is a crucial technique in providing a strong foundation for future success in mathematics, science, technology and engineering.

Why teach visual thinking?

In the primary years, visualisation is crucial when estimating, measuring and learning geometry, so that students' mathematical skills, concepts and understandings are fully developed. Visualisation techniques such as imagining, sketching and drawing, and working with maps need to be explicitly taught using materials and equipment.

How do I use *More Daily Workouts* to foster visual thinking?

Activities in *More Daily Workouts* have been designed as teacher-led discussions to promote critical thinking and verbal communication among students in learning measurement and geometry. Activities have been designed primarily to function as lesson warm-ups, in which the teacher and students discuss and develop their mathematical ideas and understandings. The activities in *More Daily Workouts* may also be used flexibly during lesson breaks or to revise previous concepts. They are laid out in inquiry style, with suggested teacher scripts, so that teachers can help students focus on concepts during discussion of various questions and problems. At the beginning of activities, students are often given a short time to investigate the idea concerned, before considering a variety of solution methods in classroom discussion and debriefing. The most efficient strategies are thus investigated and highlighted.

How do teachers assess students' use of mental imagery?

By allowing students sufficient time to develop their mental imagery and thinking, and by not intervening too soon, teachers can observe and note students' responses, which often provide a window on their thoughts. Asking students to explain their strategies by posing questions such as "Why?", "How do you know?", "Can you imagine that?" or "What would that look like?", can elicit further rich discussion.

Skills in mental imagery are thus best developed by students constructing their own knowledge from physical and mental activities related to their experience of measurement and geometry tasks. The classroom emphasis should be on investigation, imagination and discussion, so that the formation of visual images and their manipulation in the mind can be fostered.

How to use this book

More Daily Workouts is designed to offer teachers a curated collection of activities to support the teaching of mental imagery and visual thinking. The activities are intended as 5- to 10-minute introductions or warm-ups in which the teacher and students are engaged in discussion and explanation. Often the warm-ups may lead to more complete investigation of various topics by means of extension or variation activities.

Activities

Activities are organised by primary year group, strand, concept, topic and year level.
A handy thumb tab helps to identify each section – Lower, Middle and Upper. Supplementary activities that support many of the print activities in *More Daily Workouts* can also be found at www.oxfordowl.com.au.

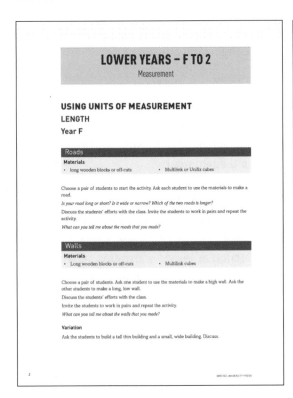

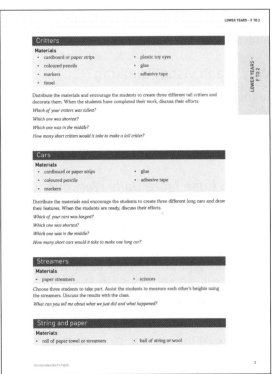

Instructions

Each activity in *More Daily Workouts* includes simple instructions and a script featuring teacher dialogue that can lead to more class discussion and consideration of the question or problem. (Sometimes the answer is also provided for quick reference.) Suggestions on how to extend or adapt activities are included where relevant.

OXFORD UNIVERSITY PRESS

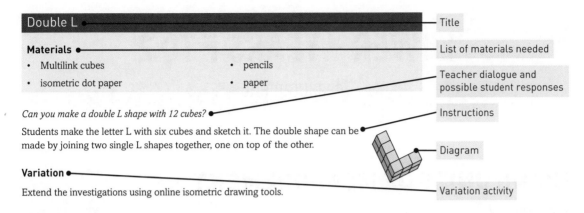

Materials

The effective teaching of measurement and geometry cannot take place without extensive use of concrete materials. Lists of materials accompany each activity in *More Daily Workouts* to allow teachers to prepare in advance. Certain classroom materials, such as paper, pencils, pens, 30 cm rulers, tape measures, metre rules, clocks, balances and masses, stopwatches, Multilink cubes, balls, grid paper, isometric dot paper and Pattern blocks, will need to be available.

In addition, teachers will need to collect a large bank of resources for use during teaching or learning activities. Some commercial resources will need to be purchased, but everyday materials can be collected in co-operation with the students and stored in the classroom. Two mobile cabinets or trolleys, one storing geometry materials and the other storing measurement materials, are a useful means of having readily accessible and transportable resources.

Online resources

The following websites offer learning experiences in digital format, and provide extension and enrichment.

- NRICH – a University of Cambridge website that provides a large variety of mathematics games and puzzles.

- Wild Maths – a rich source of games, activities and stories, also from the University of Cambridge.

- FUSE Victoria – a Department of Education and Training site that provides online activities in topics such as symmetry, tessellations, isometric drawing, non-standard measurement and telling the time.

- MAV – the Victorian mathematics association website lists many activities to support the learning of measurement and geometry.

- Wikimapia.com – an online editable map resource for investigations of location and movement.

Online searches are a feature of selected activities in *More Daily Workouts*. Search terms such as 'interactive clocks', 'domes', Australian road signs', 'maps of the world', 'learning tools for the geoboard' and 'isometric drawing tools' direct activities for teachers, provide interest and foster students' learning.

USING UNITS OF MEASUREMENT
LENGTH
Year F

Roads

Materials

- long wooden blocks or off-cuts
- Multilink or Unifix cubes

Choose a pair of students to start the activity. Ask each student to use the materials to make a road.

Is your road long or short? Is it wide or narrow? Which of the two roads is longer?

Discuss the students' efforts with the class. Invite the students to work in pairs and repeat the activity.

What can you tell me about the roads that you made?

Walls

Materials

- Long wooden blocks or off-cuts
- Multilink cubes

Choose a pair of students. Ask one student to use the materials to make a high wall. Ask the other students to make a long, low wall.

Discuss the students' efforts with the class.

Invite the students to work in pairs and repeat the activity.

What can you tell me about the walls that you made?

Variation

Ask the students to build a tall thin building and a small, wide building. Discuss.

OXFORD UNIVERSITY PRESS

Critters

Materials

- cardboard or paper strips
- coloured pencils
- markers
- tinsel
- plastic toy eyes
- glue
- adhesive tape

Distribute the materials and encourage the students to create three different tall critters and decorate them. When the students have completed their work, discuss their efforts.

Which of your critters was tallest?

Which one was shortest?

Which one was in the middle?

How many short critters would it take to make a tall critter?

Cars

Materials

- cardboard or paper strips
- coloured pencils
- markers
- glue
- adhesive tape

Distribute the materials and encourage the students to create three different long cars and draw their features. When the students are ready, discuss their efforts.

Which of your cars was longest?

Which one was shortest?

Which one was in the middle?

How many short cars would it take to make one long car?

Streamers

Materials

- paper streamers
- scissors

Choose three students to take part. Assist the students to measure each other's heights using the streamers. Discuss the results with the class.

What can you tell me about what we just did and what happened?

Year 1

Sorting lengths

Materials

- classroom objects
- relationship cards: 'is longer than', 'is shorter than', 'is the same length as'
- paper
- pencils
- coloured pencils
- markers

Choose a student and have them nominate an object in the classroom. Ask another student to select a relationship card and find another object that makes the statement true. Repeat several times and discuss the results.

Ask the students to draw and label some of the objects that they identified and discussed.

Extension

Repeat the activity using cards showing different length attributes such as: taller/shorter, thicker/thinner, higher/lower, deeper/shallower.

Door widths

Materials

- 2 doorways of differing widths
- string
- streamers or cardboard strips

Which is wider, the classroom doorway or the storeroom doorway?

How could you find out which one?

Ask the students to suggest answers and then test their predictions.

Discuss findings. Select other doorways nearby and compare their widths.

Variation

Ask the students to compare the heights of their desks and the teachers' desk.

Longer or shorter?

Materials

- pairs of objects of similar size: a pencil and a pen, a book and pencil case, a knife and a fork, pencil sharpener and an eraser, a ruler and a schoolbag

Which of these pairs of objects is longer? Which is shorter?

Are any of them the same length?

Display the objects to the students. Select students to compare the lengths of the objects by direct comparison. Explain the need for the same starting point. Discuss the findings.

Standing tall

Materials

- string
- ribbon or streamers
- scissors
- paper
- pencils

Can you find a partner who is about the same height as you?

How could we check that you are about the same height?

Discuss the different ways that the height could be measured, either by direct comparison back to back or measuring with a streamer.

Select pairs of students and measure their heights using both methods.

Discuss the results.

Make a display of the streamers, ordering them by height and labelling them with the students' names.

Year 2

Guess and step

Materials

- paper
- pencils

What is your best guess about the number of steps you would take from your chair to the classroom door? Take a good look at the distance and write down your estimate.

Select students and ask them to count the steps.

Were your estimates accurate?

Repeat the activity. Have the students estimate the numbers of steps to the whiteboard, the back wall of the room and the teachers' desk. Discuss the results. Ask the students to draw and write about what they found out.

Shortest to the longest

Materials

- string or wool
- scissors
- classroom objects

Ask the students to focus on the height of a student's chair, the width of a schoolbag, the length of a student's desk and length of the leg of a student's desk.

Which of these items will be the longest? Which will be the shortest?

Discuss the students' predictions. Invite pairs of students to measure each item using string or wool, cutting the various lengths with the scissors.

Arrange the pieces of string or wool in order. Label and display them.

Arm lengths

Materials

- string or wool
- classroom objects
- scissors

Choose a pair of students. Ask one student using the string or wool to measure the length of the other's arm from shoulder to fingertips.

Divide the class into teams and ask team members in turn to identify objects in the classroom that are shorter than the length of the arm. Check by using the string to measure the object. Each correct answer scores a point for the team.

Repeat the process for objects that are longer than the length of the arm. Total the scores and declare the winning team.

Hands and feet

Materials

- a whiteboard
- markers
- paper clips

Tell the students that they are going to investigate the lengths of their hands and feet using joined paper clips.

What is your estimate of the number of paper clips used to measure:
- *the length of your index (or pointer) finger*
- *the distance across your palm*
- *the length of your hand*
- *the distance around your wrist*
- *the length of your foot?*

Choose a pair of students in turn to measure each length. Record the results in a table on the whiteboard.

Item	Estimate	Measure
Length of index finger		
Distance across palm		
Length of hand		
Distance around wrist		
Length of foot		

Did anything unexpected happen?

Discuss the results with the students.

Desk top measures

Materials

- teachers' desk
- craft sticks
- straws
- paper
- pencils

How long is the teachers' desk?

Choose a pair of students to investigate the length of the desk using craft sticks, estimating before measuring. Select other pairs to repeat the process using straws and pencils.

How accurate were your measurements?

Which units were best for measuring? Why?

Discuss the students' responses.

Pencil count

Materials

- a set of pencils
- a set of markers
- student desks

How many pencils will fit along the length of your desk? Estimate the number and then measure your desk.

Discuss students' answers.

Now use the markers and repeat the activity.

Did you get the same result or a different result? Why were the answers different?

Discuss the difficulties of measuring with informal units, as answers vary because there is no common standard unit used.

Estimate and count

Materials

- a table
- 1 m pieces of string or 1 m cardboard strips
- pencils
- blocks
- tiles
- straws
- rulers
- school shoes
- craft sticks

If you use the same type of object, how many will fit along a 1 metre length?

Ask the students to work in pairs to estimate how many of each identical item will fit along the length.

Students record their estimates and check by measuring with the chosen items. Discuss the students' answers and compare the results.

Extension

Measure the width of the room using metre strips as a standard measuring unit.

AREA
Year F

Comparing shapes 1

Materials

• square tiles

Select three students and ask them to each arrange eight tiles to make different shapes.

In which way are these tile patterns the same? In which way are they different?

Discuss the fact that although the arrangements are different, the number of tiles and their surface space (area) are still the same.

Comparing shapes 2

Materials

• a whiteboard
• markers
• triangular pattern blocks

Draw the below shapes on the whiteboard.

Select two pairs of students and ask each pair to copy one of the arrangements using the blocks.

In which way are these tile patterns different? In which way are they the same?

Discuss the fact that although the arrangements are different, the number of tiles and their surface space (area) is still the same.

Year 1

Open and closed

Materials

- a computer
- screen
- a whiteboard
- markers

- paper
- pencils
- coloured pencils

Draw some shapes like the ones below on the whiteboard or screen.

     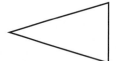

Discuss the nature of the shapes with the students.

What can you tell me about these shapes?

Ask the students to talk to the person next to them before starting a class discussion.

Focus on how the shapes can be grouped as closed and open shapes. Explain how closed shapes involve a boundary and an enclosed space within, called its area.

Encourage the students to draw open and closed shapes and colour the closed shapes.

Guess and check

Materials

- pairs of objects: an exercise book and a sheet of note paper; an A4 page and a sheet of art paper; a sheet of newspaper and a small mat

Display a pair of objects to the class.

What is the easiest way to compare the areas of these objects?

Discuss the students' suggestions. Select students to superimpose the objects, one on top of the other.

Which one has the larger area? Which one has the smaller area?

Repeat the activity for the other pairs of objects.

Did any of the objects have about the same area?

Ask the students to draw and write about what they did.

Trace and cut

Materials

- grocery packets and tins
- plates
- saucers
- saucepans
- paper
- markers

Display the collection of objects to the class. Ask the students to select two objects and trace their outlines on sheets of paper. Have them cut out the outlines, label them and directly compare them by superimposing one on top of the other.

Which outline has the larger area? Which outline has the smaller area?

Do any of the outlines have the same area? How do you know?

Variation

Ask the students to trace objects of about the same area and compare them.

Sorting areas

Materials

- classroom objects
- relationship cards: 'has a larger area than', 'has a smaller area than'
- paper
- pencils
- coloured pencils
- markers

Choose a student and have them nominate an object in the classroom. Ask another student to select a relationship card and find another object that makes the statement true. Repeat several times and discuss the results.

Ask the students to draw and label some of the objects with large and small areas.

Year 2

Cover up

Materials

- 2 student desks
- sheets of paper
- container lids or paper circles

Select a pair of adjoining desks and a number of students to assist.

Can you cover the top of the first desk with paper?

Can you cover the top of the second desk with circle shapes?

How many of each type of shape did you use?

Was all the top of the desk covered?

Which things are best for covering, the paper or circle shapes? Why?

Discuss the students' responses.

Cover the mat

Materials

- a mat or rug
- sheets of paper

Select a number of students to help with this activity.

How many sheets of paper do you think will cover the mat?

What is your best guess?

Call for students' estimates before asking the students to complete the activity.

Were your best guesses about right?

Were there any gaps or overlaps? Why?

Discuss the results.

Crafty areas

- craft sticks
- tiles or counters
- blocks

How many different closed shapes can you make with eight craft sticks?

Discuss the students' efforts, encouraging both regular and irregular shapes.

Which of your closed shapes has the largest area?

How could you find out?

Encourage the students to use blocks, tiles or counters to determine the areas of the shapes.

What did you discover?

Ask the students to draw and write about what they found out.

Cover ups 1

Materials

- container lids
- square tiles
- counters
- paper plates
- A4 pages or paper squares
- a sheet of newspaper
- an exercise book
- paper
- pencils

Divide the class into six groups and ask them to investigate the following activities.

Ask them to record their estimates and measures.

How many counters do you need to cover an exercise book?

How many square tiles do you need to cover an exercise book?

How many paper plates or round lids do you need to cover a desktop?

How many A4 pages or paper squares do you need to cover a desktop?

How many paper plates or round lids do you need to cover a sheet of newspaper?

How many A4 pages or paper squares do you need to cover a sheet of newspaper?

Have the groups report their findings to the class and discuss their responses.

Cover ups 2

Materials

- pattern blocks
- classroom objects

Can you find five objects in the classroom that can be covered by between five and ten pattern block shapes?

Select pairs of students to suggest various surfaces and check their areas using either the squares, triangles, rhombuses, trapeziums or hexagons.

Did you discover anything surprising?

Which shapes were best for measuring? Why?

Extension

Ask the students to draw irregular and regular shapes on paper and repeat the activity.

Footprints

Materials

- square tiles
- counters
- blocks
- craft sticks
- paper
- pencils
- markers

Choose a group of three students of differing heights to demonstrate this activity.

Does the tallest person in this group have the largest footprint?

How could we find out the answer to this question?

Encourage the students' responses, including direct comparison or by measuring with objects and counting them.

Ask the students to first compare the lengths of their feet from the same starting point, trace them and cover the areas with selected objects. Have them report their findings.

Was your answer true or false, or is it too early to tell?

Discuss the students' opinions.

Extension

Ask the students to investigate whether the shortest person in a group has the smallest foot.

OXFORD UNIVERSITY PRESS

Handprints

- a sheet of newspaper or butcher paper
- markers

How many handprints will cover this large sheet of paper?

Select pairs of students to fill the page with handprints. The first student places their hand flat on the paper, while the second student traces around it. Repeat the process until the page is covered.

What did you find out?

Was all the sheet covered?

Are there any gaps? Are there any overlaps?

Discuss the students' findings.

Handy shapes

Materials

- Paper
- scissors
- pencils
- markers
- counters or coins

Ask the students, working in pairs, to trace the outline of each other's hands on paper.

Have them carefully cut out the hand shapes and retain them.

Can you imagine how many counters will cover your hand shape?

Ask the students to investigate the question and report their findings.

How many hand shapes will cover a desktop?

How many hand shapes will cover a sheet of newspaper?

Ask the students to join together and investigate the questions. Discuss the results.

Handy areas

Materials

- paper
- 1 cm grid paper
- pencils
- markers

Tell the students that we are going to investigate another way to find the area of their hands. Ask the students, working in pairs, to trace the outline of each other's hands on the grid paper.

Have the students first count the whole squares to determine the area. They then can add the number of squares that are more than half a unit, also.

What did you find out? What was the area that you calculated?

Which method of measuring the area of hands was best, using counters or counting squares? Why?

Discuss the students' opinions.

VOLUME AND CAPACITY
Year F

Pack a box

Materials

- a shoe box or similar
- oranges
- tennis balls
- potatoes
- empty tins

How many oranges will fill the box?

Call for answers. Select a pair of students to fill the box.

Repeat the activity using balls, potatoes or tins. Discuss the results.

Extension

Ask the students to fill small boxes with other selected items such as cubes that will pack more tightly.

Build a house

Materials

- about 30 building blocks

Choose three pairs of students. Ask the first pair to take 10 blocks and build a tall house.

Ask the second pair to build a long house. Ask the third pair to build an unusual house.

Which house is tallest? Which house is longest?

Which house takes up the most space?

Which house did you like the most? Why?

Discuss the students' responses.

Check and tell

Materials

- 3 plastic glasses
- jug
- water

Display three glasses to the students: one full, one half-empty and one empty.

How could you describe these glasses and what is inside them?

Discuss the students' answers. Highlight the everyday language used to describe capacity, the amount that can be held by a container.

OXFORD UNIVERSITY PRESS

How many glasses?

Materials

- a wet area
- a small jug
- plastic glasses
- cups
- mugs
- water

Fill the jug and place the glasses in a line.

Ask the students to examine the amount of water in the jug very carefully.

How many glasses do you think this jug will fill?

What is your best guess?

Invite the students to fill the glasses and count the number.

Were you surprised? Did you guess correctly?

Repeat the activity for other containers.

Which holds more?

- a wet area
- a selection of containers
- a jug
- water

Choose two similar containers and display them.

Which of these containers will hold more water?

How could we find out?

Discuss the students' suggestions. Check the results by pouring from one container to another.

Can you choose two containers that will hold about the same quantity of water?

Ask the students to talk to a classmate and suggest some possibilities.

Check some of the suggestions by using a jug to fill the containers, counting the number of jugs required to fill each container. Discuss the results.

Comparing sizes

Materials

- a large ball
- 2 smaller balls
- paper
- pencils

Display the balls to the students and ask them to carefully examine them. Draw the students' attention to their size and the amount of space that they occupy.

Which ball takes up the most amount of space?

Which ball takes up the least amount of space?

Discuss the students' responses.

Ask the students to draw the balls in order of size.

Make a difference

Materials

- Multilink cubes or other cubes

Choose four or five students and give them 10 cubes each.

Can you make a shape using all 10 cubes?

Discuss the students' designs.

Repeat the activity to identify as many different arrangements as possible.

What is the same about all these different shapes you have made?

Discuss the students' answers, emphasising that the different shapes take up the same amount of space.

Year 1

Which holds the most?

Materials

- a wet area
- a selection of containers
- a jug
- water
- paper
- pencils

Choose three similar containers and display them.

Which of these containers will hold the most water?

What is the best way we could find out?

Discuss the students' suggestions. Check the results by using a jug to fill the containers, counting the number of jugs required to fill each container. Ask the students to draw the containers in order of their volume.

Jellybean jars

Materials

- 3 small jars of differing sizes
- jellybeans

Display the empty jars to the students.

Which of these jars will hold the most jellybeans?

Identify the jar and fill it with jellybeans.

How many jellybeans is this jar holding? What is your best guess?

Ask the students to estimate the number of jellybeans. Choose students to help count the number.

What did you find out about your estimates?

Discuss the results with the students.

Fill the gaps

Materials

- 5 plastic drink cups or beakers
- dry sand
- marbles
- water
- a wet area

Select five pairs of students. Tell the students that they are going to investigate packing and filling the containers and will need to make some predictions.

Ask the first pair of students to half-fill a cup with sand.

What will happen if you now add half a cup of water? Tell the class.

Ask the second pair to half-fill a cup with marbles.

What will happen if you add half a cup of sand? Tell the class.

Ask the third pair to half-fill a cup with sand.

What will happen if you add half a cup of marbles? Tell the class.

Ask the fourth pair to half-fill a cup with water.

What will happen if you add half a cup of sand? Tell the class.

Ask the final pair to half-fill a cup with marbles.

What will happen if you add half a cup of water?

Discuss the results with the students. Develop the idea that spaces between objects can be filled by other materials, both solid and liquid.

Pack and count

Materials

- 3 small cardboard boxes of different sizes
- Multilink cubes or similar

Select three pairs of students. Tell the students that they are going to investigate packing and filling the containers and will need to make some predictions.

Will a group of 10 cubes fit into your boxes? Try it.

Is there any space left in your box?

If there is space, how many more cubes do you think will fill the box to the top?

Pack them in and check.

How many cubes altogether are now in your box?

Why do you think cubes pack together very well?

Discuss the students' answers.

In for the count

Materials

- wooden bricks or similar
- shoe box
- matchbox
- eggcup
- plastic drink cup
- marbles
- counters
- paper
- pencils

Select four pairs of students. Tell the students that they are going to investigate packing and filling the containers and will need to make some predictions.

Ask the first pair of students to take a shoebox.

How many wooden bricks will pack into the shoebox?

Make your best guess and then count the bricks.

Ask the second pair to take an eggcup.

How many counters will the eggcup hold?

Make your best guess and then count the counters.

Ask the third pair to take a matchbox.

How many counters will it hold?

Make your best guess and then count the counters.

Ask the last pair to take a drink cup.

How many marbles will the drink cup hold?

Make your best guess and then count the marbles.

Discuss the students' answers and the accuracy of their predictions. Ask them to draw and write about what they have done, as an assessment tool.

Which one?

Materials

- an empty cream container
- plastic drink cup
- mug
- empty tin
- empty 600 mL drink container

Display the empty containers to the students. Ask the class to consider the volume of the containers.

How could we find out which of these containers holds the most?

Encourage the students to discuss the question with a partner. If necessary, suggest that the quickest solution is by pouring the contents of one container into the other.

What did you find out? Which container held the most? Which container held the least?

Can you now order the containers from the one that holds most to the one that holds least?

Draw the containers in order from the one that holds the most to the one that holds the least.

Year 2

Modelling cubes

Materials

- Multilink cubes
- paper
- pencils

Choose a pair of students and ask them to each take a large handful of Multilink cubes.

Can you build a model with your cubes?

When the building is complete, ask the class to compare and talk about the models.

Which model takes up the most space?

How could we find out?

Encourage the students to count the cubes to check the model that has the larger volume.

Ask the students to record the results.

_____ 's model used _____ cubes.

_____ 's model used _____ cubes.

_____ 's model took up the most space.

Drawing cubes

Materials

- Multilink cubes
- isometric dot paper
- pencils

Display a single cube to the students.

How could we draw this cube on the dot paper?

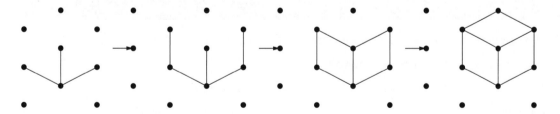

Demonstrate the correct way of drawing the cube with its front faces orientated at 300-degree angles. Check the students' drawings.

Extension

When the students are ready, ask them to draw two cubes joined together and finally three cubes. Discuss the different arrangements with the students. Display the students' work.

Tetracubes

Materials

- Multilink cubes
- isometric dot paper
- pencils

How many ways can you join four Multilink cubes together?

Display a single tetracube and invite the students to work alone or in pairs to investigate other tetracubes. There are eight possible combinations.

What is the same about each model?

Do they each take up the same amount of space?

How many different tetracubes did you discover?

Discuss the different arrangements with the students.

Variation

Use Base 10 minis or other cubes to make tetracubes.

Drawing tetracubes

Materials

- tetracube models from previous activity
- isometric dot paper
- pencils
- coloured pencils

Can you draw your favourite tetracube on isometric dot paper?

Discuss the students' efforts and display the students' drawings.

Extension

Encourage the students to draw other selected tetracubes.

Volume quiz

Ask the students the following questions to assess their understandings.

Discuss their responses. Answers may vary.

Which object has a greater volume?

A balloon or a watermelon?

A tennis ball or golf ball?

A soccer ball or a softball?

A tennis ball or a cricket ball?

A pillow or a cushion?

A can of drink or a carton of juice?

A Duplo block or a matchbox?

A glue sticker or a marker?

A matchbox or an eraser?

OXFORD UNIVERSITY PRESS

MASS
Year F

Eyes wide shut

Materials

- a tennis ball
- a full pencil case
- an orange
- an eye mask

Call for a volunteer. Wearing a blindfold, ask them to feel the objects one by one.

Which of these objects is heavy? Which object is light?

Discuss the students' findings with the class.

Can you put the objects in order from lightest mass to heaviest mass?

Extension

Repeat the activity with other sets of objects.

Drag it along

Materials

- ball of string
- a thick library book
- backpack
- a box of blocks
- a medicine ball

Select a group of three or four students. Help them tie a string around each of the objects.

Ask the students to take turns pulling the object along the floor.

Which object is heaviest? Which object is lightest? How do you know?

Discuss the students' findings and the reasons for their judgements.

Extension

Repeat the activity for other objects.

Pass the parcel

Materials

- a selection of 4 or 5 wrapped parcels: some small and heavy, others large and light

Display the parcels to the students.

Which parcel do you think will be the heaviest?

Which one will be the lightest?

Discuss the answers with the students and arrange the parcels in an agreed order from heaviest to lightest. Invite the students to handle the parcels and check the class estimates.

What did you discover about your estimates?

If you wish, allow the students to unwrap the parcels and check their contents.

Develop the idea that the volume of an object does not necessarily predict its mass.

Heavy or light?

Materials

- 2 hoops or 2 sheets of butcher paper
- label cards: 'heavy' and 'light'
- assorted classroom objects
- paper
- pencils
- coloured pencils

Indicate a selection of classroom objects.

Which things are heavy? Which things are light?

Ask the students to suggest answers. Invite the students to place the objects in two hoops or on two sheets of butcher paper labelled heavy or light. Check their selections.

Can you draw your favourite heavy objects?

Can you draw your favourite light objects?

Display the students' work.

Heavy rock

Materials

- a whiteboard
- markers
- a rock or large pebble
- classroom objects

Can you name objects in the classroom that are lighter than this rock?

Call for answers. Invite selected students to test the students' suggestions by hefting.

To do this, the students hold one object in each hand, raise them up and down and decide which is lighter and which is heavier. Make sure the students understand that this is 'hefting' and become familiar with the word.

Record the results on the whiteboard.

_____ found that the _____ is lighter than the rock.

OXFORD UNIVERSITY PRESS

Order of rocks

Materials

- 5 rocks or large pebbles

Which of the rocks is the heaviest? Which of the rocks is the lightest?

Can you find two rocks that have similar mass?

Select students to investigate the questions by hefting. Discuss the findings.

Variation

Use assorted library books instead of rocks.

Year 1

See-saw

Materials

- a metre stick
- block of wood
- book
- rock
- a full pencil case
- stapler
- calculator

Can you make a see-saw with the block and the metre stick?

Can you predict which is the heaviest object using the see-saw?

Discuss the students' choices and the reasons behind them. Select some students to test their predictions.

Can you arrange the objects in order from lightest to heaviest?

Discuss.

Balance it up

Materials

- a bucket balance or equal-arm balance
- pairs of classroom objects: a rock and ball, a pencil and an eraser, a paintbrush and a cup, a calculator and a stapler, a marble and a Multilink cube

Which of the two objects feels heavier? How could you check your answer?

Select students to investigate the mass of each pair of objects using the balance.

Ask them to predict the heavier of the two objects before measuring them.

What did you find out about your estimates?

Talk about the students' findings.

Levelling out

Materials

- a bucket balance or equal-arm balance
- classroom objects

Can you name some things that will balance each other exactly?

Ask the students for their suggestions. Invite them to check their choices by using the balance. Discuss the results.

Equal masses

Materials

- a bucket balance or equal-arm balance
- coins
- Multilink cubes
- marbles
- a stapler
- paper clips
- counters
- craft sticks
- an eraser

Invite a pair of students to investigate each of the following questions.

How many coins will balance five Multilink cubes?

How many marbles will balance a stapler?

How many paper clips will balance five counters?

How many counters will balance a Multilink cube?

How many craft sticks will balance an eraser?

Did you find anything unusual or surprising?

Discuss the results with the class.

Extension

Provide opportunities for the students to investigate equal masses during play.

Balancing crayons

Materials

- a bucket balance or equal-arm balance
- a packet of markers or crayons
- Multilink cubes
- coins
- craft sticks
- pencils
- marbles

How many of these small objects will balance a packet of markers?

Ask the students for their suggestions. Invite them to check their choices by using the balance. Discuss the results.

OXFORD UNIVERSITY PRESS

Measure then estimate

Materials

- a bucket balance or equal-arm balance
- a pencil
- a marker
- an eraser
- counters
- a tennis ball

How many counters do you think it will take to balance: a pencil, a marker and an eraser?

Select students to investigate the answers for each of the objects.

Record the results and discuss them with the students.

What do you think would be the mass of a tennis ball measured using counters?

Ask the students for their estimates before selecting students to determine the answer.

Year 2

Conserving mass

- a bucket balance or equal-arm balance, playdough or plasticine, several pebbles

Make a ball of playdough that weighs the same as several pebbles. Confirm with the students that the masses are equal.

If we change the ball of playdough into a sausage shape, will it still balance the pebbles?

Ask the students to make their predictions. Check the result by measuring. Discuss the result.

If we divide the playdough into two pieces, will it still balance the pebbles?

Invite a student to divide the playdough in half and carry out the activity. Discuss the results.

Highlight the fact that although the shape of the playdough has changed, its mass, or amount of substance, remains the same.

Match an object

Materials

- a bucket balance or equal-arm balance
- Multilink cubes
- classroom objects
- paper
- pencils

Can you name some things that will have about the same mass as 10 Multilink cubes?

Ask students for their suggestions. Invite them to check their choices by using the balance.

Discuss the results. Record the results.

Ten cubes have the same mass as one _____.

Extension

Use 10 pebbles, marbles, nuts, bolts or washers and repeat the activity.

Compare the two

- a bucket balance or equal-arm balance, Multilink cubes, craft sticks

Which weighs more: 20 craft sticks or 20 Multilink cubes?

Call for estimates from the students. Discuss their responses.

Invite a pair of students to investigate the answer by measuring.

Which weighed more?

How accurate were your estimates?

Discuss.

Extension

Repeat the activity for other numbers of objects suggested by the students.

Cube balance

- a bucket balance or equal-arm balance, Multilink cubes, full pencil case, whiteboard eraser, softball, drink bottle, cup, mug

Display the objects to the class.

How many cubes will balance each of these objects?

Call for estimates from the students. Discuss their responses.

Invite the students to investigate the answer by measuring.

What did you discover about your estimates?

Which object is the heaviest?

Which object is the lightest?

How many more?

Materials

- a bucket balance or equal-arm balance
- a roll of tape
- a calculator
- Multilink cubes

Place a roll of tape on one balance pan and a calculator on the other.

How many cubes do you think have to be added to one pan to balance the roll of tape and the calculator?

Call for answers. Discuss the students' responses.

Ask the students to suggest other pairs of objects in the room whose mass can be investigated.

OXFORD UNIVERSITY PRESS

Needing a standard

Materials

- an apple or orange
- Multilink cubes
- coins
- pencils

Select students to measure the mass of the apple using the different materials.

Are the results the same? Why not?

Can you think of any way to make the results the same?

Discuss the students' suggestions. Introduce the idea, if necessary, that accurate measurement is only possible if standard units are used.

TIME

Year F

Long or short?

Materials

- paper
- pencils
- coloured pencils

During the day, which activities take a long amount of time? Which activities take a short amount of time? Which activities go on forever?

Ask the students to draw the different activities. Discuss and display their work.

Follow the bouncing balls

Materials

- a tennis ball
- rubber ball
- a basketball

Select three students and give them each a ball.

If you drop the balls at the same time, which ball will bounce for the longest time?

Invite the students to predict the answer. Carry out the activity and discuss the results.

Days of the week

Materials

- days of the week flashcards
- paper
- pencils
- coloured pencils
- markers

Ask a pair of students to arrange the cards in weekly order. Select students to choose cards representing today, yesterday and tomorrow.

Can you draw something that you did today?

Can you draw something that you did yesterday?

Can you draw something that you will do tomorrow?

Discuss the students' efforts and display them.

Paper chains

Materials

- 5 paper strips of a single colour labelled Monday–Friday
- 2 paper strips of another colour labelled Saturday and Sunday
- stapler or glue stick
- marker

Make a paper chain of the days of the week. Invite the students to answer some questions, referring to the chain.

What day is it today?

What day is it tomorrow?

What day was it yesterday?

What day comes before _____?

What day comes after _____?

Why are two loops a different colour?

How many days are there until the weekend?

How many days till _____?

How many days are there in 1 week?

How many days are there in 2 weeks?

What is another name for a period of 2 weeks?

Encourage the students to make their own paper chains.

Special times

Materials

- paper chains from previous activity
- paper
- pencils
- coloured pencils

Talk about special times during the week, such as a training session or match for a sport, or a music lesson, a regular walk with family or a play session with a friend.

Can you draw a picture to show a special thing that you do on each day of the week?

Display the students' drawings.

Draw the clock

Materials

- classroom clock or clock model
- pencils
- paper

Close your eyes. Imagine a clock face.

What does it look like? Can you draw it from memory?

Ask the students to draw a clock face and its hands. Discuss the different features of their drawings. Compare the students' drawings with a classroom clock or clock model.

Were your drawings accurate? Were the numerals in the correct positions?

Where did you place the hands?

Are the hands different lengths? Why?

Can you make your clock show a time?

What time is it?

Discuss the students' responses.

Year 1

Colour the clock

Materials

- a whiteboard
- markers
- a worksheet showing 12 30-degree segments
- pencils
- coloured pencils

Draw a clock face on the whiteboard showing 12 30-degree segments.

Invite a student to suggest a particular hour.

Where do I draw the hour hand?

Where do I draw the minute hand?

Is this correct?

Colour in the segment between the hour hand and minute hand, representing the hour. Ask the students to select an hour and repeat the activity using their worksheets.

Discuss the results.

Hands on hours

Materials

- a computer
- screen
- a whiteboard
- markers

- worksheets showing multiple blank analogue clock faces
- pencils
- paper

Download an interactive clock from the Internet or draw a blank analogue clock face on the whiteboard.

Today we're going to tell the time on the hour.

What is the position of the hands at 11 o'clock?

Encourage the students' responses and check their answers.

Point out that the hour hand should be exactly on 11 and the minute hand should be on 12. (Try to avoid saying 'big hand' and 'little hand', as they do not convey precise meaning). Distribute the pages showing a blank analogue clock face.

Can you choose another hour and draw the hands on the clock face?

Discuss the students' selections with the class.

Match the times

Materials

- a model clock or paper plate clock

Select different students to use the clock to model the hour that they wake up, go to school, eat lunch, have dinner and go to bed.

Ask the students to select other activities during the day and model their times.

The seasons

Materials

- photographs or illustrations of different seasons
- pattern blocks

Develop a picture and talk about the seasons and the changes in the natural environment that occur. Ask the students to use the pattern blocks to create designs representing events or items associated with different seasons such as:

- a sandcastle at the beach in summer
- a fallen leaf in autumn
- a scarf in winter
- and a flower in spring.

Photograph and display the students' designs.

Extension

Encourage the students to represent other events or items using various media.

Year 2

Hands on half hours

Materials

- a computer
- screen
- a whiteboard
- markers

- worksheets showing multiple blank analogue clock faces
- pencils
- paper

Download an interactive clock from the Internet or draw a blank analogue clock face on the whiteboard.

Today we're going to tell the time on the half hour.

What is the position of the hands at half past 11 o'clock??

Encourage the students' responses and check their answers.

Point out that the hour hand should be exactly halfway between 11 and 12 and the minute hand should be on 6. Distribute the pages showing a blank analogue clock face.

Can you choose another half past the hour and draw the hands on the clock face?

Discuss the students' selections with the class.

Check answers. Ensure that the students have drawn the minute hand so that it divides the clock face in half and that the hour hand is exactly halfway between the 2 hours in question. Accept the description of 30 minutes past the hour.

Birthday graph

Materials

- 12 balls of plasticine or playdough
- label cards for the months of the year
- toothpicks or stick counters

Arrange the 12 balls of plasticine in a row. Label each ball to show a month of the year. Invite the students to place toothpicks or sticks in their birthday months.

Which month had the most birthdays?

Which month had the least birthdays?

Did any month have the same number of birthdays?

Were there any months with no birthdays?

How many students had birthdays in spring?

How many students had birthdays in summer?

How many students had birthdays in autumn?

How many students had birthdays in winter?

Discuss the results with the students.

Calendar counting

Materials

- a calendar
- paper
- pencils

Use a calendar and ask the students to assist with answering the following questions.

How many days are there in each month?

How many weekends are there in this month?

How many days are there in this month?

What is the date of the first Saturday of this month?

How many school days are there in this month?

How many school days until the school holidays?

How many weeks of school in Term 2?

How many school days in Term 2?

What are the dates of Easter?

What are the dates of Ramadan?

What are the dates of Chinese New Year?

Discuss the answers. Encourage the students to make up questions of their own and answer them.

OXFORD UNIVERSITY PRESS

Calendar days

Materials

*   multiple copies of a particular calendar month
*   pencils

Provide the students with a copy of the calendar month. Encourage the students to say the days of the week as you point to them. Point to the numbers and count them.

Can you see any repeating patterns in your month?

Discuss horizontal, vertical and diagonal patterns discovered by the students.

What is the pattern formed by Mondays?

On which day is the first day of the month? On which day is the last day?

How many Wednesdays and Fridays are there in your month?

What day of the week will it be on the 2nd, 9th, 16th and 28th?

What date is the second Tuesday? How many weekdays are there?

How many weekends are there?

Discuss the answers. Encourage the students to make up some questions of their own and answer them.

LOWER – YEARS F TO 2
Shape

SHAPE
THREE-DIMENSIONAL OBJECTS
Year F

Coloured towers

Materials

- Multilink cubes or similar
- a small box or towel

Invite four students to construct four short towers with the cubes, with each tower made up of a different colour. Display the towers to the class and then cover them.

Can you remember the colour of each tower in order?

Call for answers. Discuss the students' suggestions.

Rearrange the order and repeat the activity several times.

Extension

Increase the number of different-coloured towers used.

What is missing?

Materials

- tennis ball
- eraser
- sharpener
- block
- marker
- 2 L ice cream container with lid

Display the items to the students. Select a pair of students and ask the first student to close their eyes. The second student then selects an object, places it in the container and seals it. The first student then guesses what is missing from the group of objects, that is, what is in the container. The second student then takes their turn. Repeat the activity with other pairs of students.

OXFORD UNIVERSITY PRESS

Copy my model

Materials

- Multilink cubes or similar

Select a pair of students and ask them to build a model using a small number of cubes.

Ask the remaining students to work in pairs and build copies of the first model.

Extension

Repeat the activity over time, increasing the number of cubes used.

Variation

Students work in pairs. The first student builds a model, and the partner then copies it.

Crash it

Materials

- Multilink cubes or similar

Build a small tower using a selection of coloured cubes. Pull the tower apart quickly and ask the students to help you reconstruct the tower from memory.

Ask the class to work in pairs and repeat the activity.

Extension

Repeat the activity over time, increasing the number of cubes used.

Copy from memory

Materials

- Multilink cubes or similar
- towel

Ask the students to work with a partner. The first student makes a model with a small number of cubes and then covers it up.

Do you remember what the model looks like?

Can you make it again with the blocks?

Students then uncover the second model and compare the two.

What did you find out when you compared the two models?

Discuss the students' findings.

Block walls

Materials

- Multilink cubes or similar
- paper
- pencils

Can you build a long wall with the cubes?

Ask the students to work in pairs to make their models.

Is your wall straight or curved?

How many blocks high is your wall? How many blocks long is it?

Are there any doors or windows in your wall?

Ask the students to draw their walls. Discuss and display the students' drawings.

Staircases

Materials

- Multilink cubes or similar
- paper
- pencils

Can you build a staircase with the cubes?

Ask the students to work in pairs to make their models.

How many blocks high is your staircase? How many blocks wide is it?

How many steps are there in your staircase?

Ask the students to draw their staircases. Discuss and display the students' drawings.

Mystery objects

Materials

- a shopping bag or cardboard box
- a selection of objects such as a tennis ball, a block, timber offcut, a tin, plastic bottle, an eraser, a pencil sharpener, a cup

Place the objects in the bag, out of sight of the students. Invite a student to reach inside the bag, select one of the objects and manipulate it.

Ask the other students to pose a series of questions that can only be responded to by answering yes or no, such as:

Is it rough? Is it smooth? Is there a hole in it? Has it got an inside and an outside? Does it have a flat surface? Can you feel its sides? Is it curved? Does it have corners?

Invite the students in turn to ask a series of questions, as they gradually form a mental image of the shape and attempt to discover it, step by step, instead of quickly guessing it. Repeat with other objects.

Year 1

Toothpick designs

Materials

- toothpicks
- Blu Tack or playdough
- paper
- pencils

How many different structures can you make with toothpicks?

Encourage the students to create three-dimensional (3D) frameworks using toothpicks joined by small pieces of Blu Tack. Discuss the different designs made by the students.

Which is your favourite design? What does it look like?

Discuss the students' ideas. Ask the students to copy their designs on paper. Display their work.

Toothpick shapes

Materials

- toothpicks
- Blu Tack or playdough
- paper
- pencils

Can you make some boxes with toothpicks?

Encourage the students to make a range of shapes. Ask the students to copy their designs on paper. Display their work.

Magnetic 3D shapes

Materials

- Geomag magnetic shapes
- camera

Can you make some different 3D shapes with the Geomag shapes?

Encourage the students to make a range of shapes. Take photographs of the students' models and display them.

Inside shape hunt

Materials

- 3 hoops
- cardboard labels
- a collection of everyday objects such as balls, boxes, packets, tins, cartons, bottles, tubes, blocks and wooden offcuts

Ask the students to identify everyday items whose shape resembles the basic geometric shapes.

Can you find objects that are shaped like prisms, cylinders and spheres?

Ask the students to place each item in a different hoop. Place labels next to the different hoops: 'shapes that look like prisms', 'shapes that look like cylinders' and 'shapes that look like spheres'. Discuss the results with the students.

Outside shape hunt

Materials

- outdoor environment
- cameras

Take the students outside the classroom to investigate the design of some of the school buildings. Ask the students to describe the basic geometric shapes they see.

What kind of shapes can you see when you look at this building?

The students will be first inclined to offer suggestions based on their perceptions of two-dimensional (2D) shapes such as triangles, rectangles and circles. This should be discussed before drawing their attention to the idea of shape in three dimensions also.

Can you find objects that are shaped like prisms, cylinders and spheres?

Photograph the shapes pointed out by the students and make a classroom display of the different groups.

Moving shapes

Materials

- a collection of everyday objects such as balls, boxes, packets, tins, cartons, bottles, tubes, blocks and wooden offcuts, paper, pencils

Display the objects to the students.

Which shapes roll? Which shapes slide? Which shapes stack?

Ask the students to draw an example of each type of shape.

Pull it apart

Materials

- empty toothpaste, cereal, tissue or other grocery boxes
- a letter opener

Can you describe how these boxes are made?

Discuss the students' suggestions.

What would this box look like if it was unfolded and carefully opened out?

What shapes would you see?

Use the letter opener to release the sides of the box.

Were your ideas accurate?

Ask the students to draw the shape of the unfolded box.

OXFORD UNIVERSITY PRESS

Playground models

Materials

- Multilink cubes or similar
- paper
- pencils

Take the students for a walk in the school grounds. Ask them to sketch one of the buildings.

Return to the classroom.

Can you use the blocks to make a model of the building, using your sketch to help?

Take photos of the models made by the children and display them.

Town models

Materials

- Multilink cubes or similar
- paper
- pencils

Select small groups of students to investigate this activity.

Can you build a model of a town?

Allow time for the students to complete their designs before discussing their features.

Imagine you are a bird flying over the town and looking down.

Can you draw a plan of what you would see?

Discuss the students' plans.

Now imagine yourself as a bug looking straight ahead at the town.

Can you draw a plan of what you would see?

Display the students' drawings.

Year 2

Faces, edges and corners 1

Materials

- a collection of wooden offcuts or large blocks
- grocery packets
- pencils
- paper

Select small groups of students to investigate this activity.

Ask the students to select an item and run their fingers over the faces.

What does this object feel like? Is it smooth, rough, cold or flat?

What are the shapes of its faces?

Discuss the students' answers.

Indicate the edge of the block to the students.

What are the parts of the block called where these two faces meet?

Introduce the term 'edges' to describe those parts of the block.

Indicate the corner of the block to the students.

What are the parts of the block called where three faces meet?

Introduce the term 'corners' to describe those parts of the block.

The term 'vertices' (singular 'vertex') may also be used.

Ask the students to draw a picture of their object. Have them investigate and record how many faces, edges and corners it has.

Faces, edges and corners 2

Materials

- cans
- tubes
- Cornetto wrappers
- balls

Select small groups of students to investigate this activity.

Ask the students to select an item and run their fingers over the faces.

What does this object feel like?

What are the shapes of its faces?

Discuss the students' answers.

Indicate a cylinder-shaped object to the students.

What is different about the faces of this object?

Discuss the fact that objects may have curved and flat faces.

Repeat the activity for a cone-shaped object.

What is different about the curved shape of this object?

Indicate and name the vertex of the cone-shaped object.

What about this ball? What is special about its shape?

Review the shapes and their different characteristics. Ask the students to draw a picture of the objects. Have them investigate and record how many faces, edges and corners they have.

Hunting faces, edges and corners

Materials

- outdoor environment
- paper
- pencils

Take the students outside the classroom to investigate the design of some of the school buildings.

OXFORD UNIVERSITY PRESS

Can you identify faces on parts of this building? Can you point out any edges?

What about corners?

Discuss the students' suggestions. Ask the students to draw some of the examples of faces, edges and corners that they identified.

Imagine and cut 1

Materials

- a cube made from modelling clay or playdough
- fishing line or strong thread

Select a pair of students for this activity.

Look at this cube.

Can you imagine how you could cut it to show a square face?

Encourage the students to discuss their ideas.

Try it and see. Were your predictions correct?

Can you cut the cube to show a rectangular face? A triangular face?

Discuss the results with the students.

Imagine and cut 2

Materials

- a cylinder made from modelling clay or playdough
- fishing line or strong thread

Select a pair of students for this activity.

Look at this cylinder.

Can you imagine how you could cut it to make a circular face?

Encourage the students to discuss their ideas.

Try it and see. Were your predictions correct?

Can you cut the cube to make an oval face? A rectangular face?

Discuss the results with the students.

Imagine and cut 3

Materials

- a cone made from modelling clay or playdough
- fishing line or strong thread

Select a pair of students for this activity.

Look at this cone.

Can you imagine how you could cut it to make a circle?

Encourage the students to discuss their ideas.

Try it and see. Were your predictions correct?

Can you cut the cube to make an oval? A rectangle?

Discuss the results with the students.

Mystery models

Materials

- geometric models of prisms, pyramids, cylinders, cones and spheres

Display the various geometric models to the students.

Do you know the names given to these shapes?

Review the various names given to the shapes, including cube, prism, pyramid, cylinder, cone and sphere. Hold up a particular model and point out its faces, edges and corners.

Place the objects in the bag, out of sight of the students. Invite a student to reach inside the bag, select one of the objects and manipulate it.

Ask the other students to ask successive yes/no questions in order to identify it.

Does it have flat surfaces? Does it have a lot of faces? Is it curved? Is it pointy? Does it have any corners? Does it have a flat face and a curved face?

Encourage the students to pose their questions and work out their answers gradually, without the need for guessing.

Frame it

Materials

- toothpicks
- Blu Tack or playdough
- paper
- pencils

Can you design a frame for a small tent?

Encourage the students to use the materials to investigate different frame designs.

How many edges are there on your frame?

How many corners are there?

Can you draw the shape of the floor of your tent?

Discuss the students' efforts. Encourage them to make frameworks of other geometric shapes.

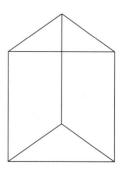

OXFORD UNIVERSITY PRESS

Domes

Materials

- a computer
- screen
- online search tools

Search the term 'domes' online.

Display the images to the students and discuss the meaning of the term and how domes have been utilised by different cultures throughout history.

Focus on the images of the dome-building kits and explain to the students how a small-scale dome can be constructed.

What is the main shape used by the dome kit?

Which geometric shape does the finished dome remind you of?

Discuss the term 'hemisphere' with the students.

TWO-DIMENSIONAL SHAPES

Year F

Trace it

Materials

- boxes
- blocks
- timber offcuts
- cups
- books
- butcher paper

Ask the students to work in pairs for this activity. The first student chooses an object, and the second student holds it tightly on top of the paper as the first student traces around it.

Repeat the activity several times.

What did you find? Are the shapes the same or different?

What are the names you could give to the shapes you made?

Develop the idea that 2D shapes originate as the faces of 3D objects.

Sort it

Materials

- pattern blocks

Set out a collection of pattern blocks on the table.

How could we sort these blocks?

Choose individual students to sort the blocks by colour, shape or other criteria suggested by the students. Discuss the different arrangements.

Who can make the tallest tower with the blocks?

Shape designs

Materials

- pattern blocks
- paper
- pencils
- coloured pencils
- markers

Display a collection of pattern blocks to the students.

Can you make a house? A flower? A snake? A dog? A bird? A person? A fish?

Discuss the students' efforts. Ask them to draw their favourite design.

Extension

Encourage the students to create designs of their own.

Make a pattern

Materials

- pattern blocks
- camera

Display a collection of pattern blocks to the students.

Can you name the shapes of the different pattern blocks?

Discuss the different shapes, squares, triangles, trapeziums, hexagons and two different diamonds or rhombuses.

What do you think is special about these blocks?

Discuss the way the blocks fit together without creating any gaps or overlaps.

Can you make pattern with the blocks?

What does it look like? Is it an animal? A person? A flower?

Discuss the students' patterns. Photograph and display them.

Extension

Allow the students to play freely with the blocks outside of normal lesson time.

Variation

Prepare pattern block shape templates and encourage the students to cover them with blocks.

Shape patterns

Materials

- pattern blocks
- camera

Display a collection of pattern blocks to the students.

How many patterns can you make with one type of block?

OXFORD UNIVERSITY PRESS

How many patterns can you make with two types of blocks?

How many patterns can you make with three types of blocks?

Discuss the students' patterns. Photograph and display them.

Block copies

Materials

- pattern blocks
- paper
- pencils

Ask the students to choose three different blocks and join them together to make a shape.

Can you draw a picture of your shape?

When the students have completed their drawings, ask them to exchange their drawings with each other and build one another's shapes.

Extension

Encourage the students to make more complex shapes with the blocks and copy them onto paper.

Upside down

Materials

- pattern blocks or similar

Select a pair of students. Ask the first student to make a simple design with the blocks.

Ask the second student to copy the design upside down. Discuss the designs.

Ask the remaining students to work in pairs and repeat the activity.

Road signs

Materials

- a computer
- screen
- online search tools

Search the term 'Australian road signs' online. Display the images to the students.

Ask the students to describe the different shapes that the signs are based on.

Which shapes are triangular? Rectangular? Circular?

Is the diamond road shape really a diamond, or is it a square standing on its corner?

Talk with the students about the meanings that the shapes signify.

Human shapes

Materials

- a large group of students
- camera

Find an open space outside the classroom. Ask the students to hold hands with the people in their group.

Can you arrange yourselves to make a square? A triangle? A rectangle?

What other shapes can you make?

Take photos of the shapes made by the children and display them.

Rope shapes

Materials

- a small group of students
- a rope of about 10 m length joined at each end

Find an open space outside the classroom.

Can you step inside the rope and make some shapes?

Can you make a square? A triangle? A rectangle? A circle? A diamond? A hexagon?

What other shapes can you make?

Take photos of the shapes made by the children and display them.

Board copies

Materials

- a whiteboard
- markers
- paper
- pencils

Draw the following designs on the whiteboard.

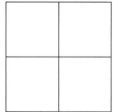 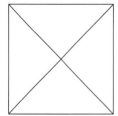  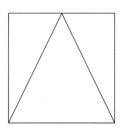

Ask the students to study the designs and copy them on paper. Discuss their efforts.

How many different shapes can you see? What kind of shapes are they?

Cut the toast

Materials

- paper
- pencils

What is the best shape to cut a square of toast or sandwiches into?

Call for answers and discuss the students' suggestions.

Do you prefer triangles, squares or rectangles?

Ask the students to draw the various shapes discussed.

Year 1

Numbers upside down

Materials

- a whiteboard
- markers

Write the following numbers on the whiteboard: **11, 16, 19, 66, 99, 101, 116, 161, 191**

Can you imagine what these numbers would look like if they were turned upside down?

Demonstrate the answers to the students and discuss them.

Variation

Ask the students to enter the numbers into a calculator, imagine what they would look like and then turn the calculator upside down.

Imagine this

Materials

- a whiteboard
- markers
- paper
- pencils

Copy this diagram onto the whiteboard.

This is a sheet of paper that has been folded in half and pieces have been cut out of it.

Can you imagine what it would look like when it is unfolded?

How would you describe the shape when it is unfolded?

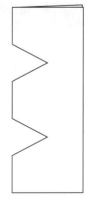

Talk about the shapes predicted by the students. Ask them to draw the unfolded shape. Discuss the students' drawings. Cut out the shape and reveal it to the students.

Did your drawings look like the cut-out shape?

Craft stick shapes 1

Materials

- craft sticks
- paper
- pencils

How many different flat shapes can you make with craft sticks?

Are your flat shapes open shapes or closed shapes?

Discuss the different shapes made by the students. Ask the students to copy their designs onto the paper.

How many different sides do your shapes have?

Can you point out their angles?

How many different angles do your shapes have?

Discuss the students' responses.

Geostrip shapes 1

Materials

- geostrips or identical cardboard strips
- split pin paper fasteners

Invite four students to investigate this activity.

Give 3 strips to the first student, 4 strips to the second, 5 strips to the third and 6 strips to the fourth student. Ask them to join their strips with the paper fasteners.

What is the name given to this three-sided shape? This four-sided shape? This five-sided shape? This six-sided shape?

Ask the students to manipulate their shapes.

What happens when you move the sides of these shapes?

Which shape does not change when you push its sides?

How could you make four-sided shapes more rigid?

Craft stick shapes 2

Materials

- a large amount of craft sticks
- glue
- paper or cardboard

Select two students to commence this activity.

Can you make a triangle with craft sticks?

Can you make a diamond (rhombus) with four craft sticks?

Display the students' efforts. Distribute the craft sticks to the rest of the class.

Can you make your own shape with a large number of craft sticks?

Can you see small triangles in your shape?

What about larger triangles?

Are there any diamonds (rhombuses)?

Can you see a trapezium or a hexagon?

Glue a selection of craft stick shapes to a sheet of paper or cardboard to make a display. Label the shapes produced by the students.

Geostrip shapes 2

Materials

- geostrips or thin cardboard strips
- split pins or paper fasteners

Select a group of students to demonstrate this activity.

Use the strips to make a triangle, a square, a rectangle and a diamond.

Can you make a hexagon?

Can you change your hexagon into a triangle? A rectangle? A shape with five sides?

Do you know the name for a five-sided shape?

Ask the students to draw a picture of what they did.

Magnetic 2D shapes

Materials

- Geomag magnetic shapes
- camera

Can you make some 2D shapes with the Geomag shapes?

Encourage the students to make a range of shapes.

Ask them to count their sides and angles.

Take photographs of the students' models and display them.

Geoboard shapes

Materials

- geoboards
- coloured rubber bands
- square dot paper
- pencils
- coloured pencils
- markers

Ask the students to make some shapes with the geoboards and coloured rubber bands.

How many different triangles can you make on your geoboard?

Monitor the different triangles made by the students, including equilateral, isosceles and scalene triangles, so that a broad range is considered.

How many different four-sided shapes (quadrilaterals) can you make on your geoboard?

Review the squares, trapeziums, parallelograms, diamonds or rhombuses constructed by the students. Ask them to copy their favourite shape onto the square dot paper.

Dotty shapes

Materials

- square dot paper
- pencils
- rulers

Give each student a sheet of square dot paper.

Can you join any three dots with straight lines? What shape did you make?

What about four dots? What shape did you make?

Discuss the students' efforts. Review triangles, squares, trapeziums, parallelograms, diamonds or rhombuses drawn by the students.

What shape can you make by joining six dots?

Encourage the students to draw a range of hexagons, having sides of different lengths rather than just the regular shape.

City shapes

Materials

- a computer
- screen
- online search tools

Search the term 'City Skyline Silhouette Images' online. Select an image and display it to the students.

Ask the students to describe the different shapes that they see.

Do you see both straight-sided shapes and curved shapes?

Which type of shape is more common?

Discuss the students' answers.

Coloured shapes

Materials

- square dot paper
- pencils
- rulers
- coloured pencils
- markers

Give each student a sheet of square dot paper.

Can you draw a green square? A thin blue rectangle? A red hexagon? A yellow star?

Which other shapes can you make? Colour them.

Talk about the students' designs.

OXFORD UNIVERSITY PRESS

Punching holes

Materials

- a hole punch
- A4 paper sheets

If we use the hole punch on a single sheet of paper, how many holes will it make?

Call for answers. Ask a student to carry out the activity and check the result.

What if we fold the paper in half, how many holes will there be then?

Discuss the students' predictions. Ask a student to carry out the activity and check the result.

Were your predictions correct?

What about we fold the paper in quarters? What would happen then?

Discuss the students' predictions. Ask a student to carry out the activity and check the result.

What happened? Can you see a pattern forming here?

Discuss the results with the students.

Year 2

Imagine this

Materials

- a whiteboard
- markers
- paper
- pencils

Copy this diagram onto the whiteboard.

This is a sheet of paper that has been folded in half and pieces have been cut out of it.

Can you imagine what it would look like when it is unfolded?

How would you describe the shape when it is unfolded?

Talk about the shapes predicted by the students. Ask them to draw the unfolded shape. Discuss the students' drawings. Cut out the shape and reveal it to the students.

Did your drawings look like the cut-out shape?

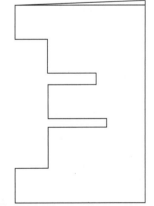

On the line

Materials

- pattern blocks
- ruler

Lay a ruler vertically on the table. Invite a student to use the blocks to create a design along its side. Ask another student to copy the design on the other side of the ruler.

Display the arrangement to the class.

What can you tell me about these designs?

What do they look like?

What do you think looks the same about them?

Do the shapes match on each side of the ruler?

Choose another pair of students and repeat the activity.

Discuss the students' efforts. Introduce the terms 'mirror image' and 'symmetry' to describe the patterns.

Block symmetry 1

Materials

- a whiteboard or computer
- screen
- pattern blocks

Copy the following block arrangement on the whiteboard or screen.

Challenge the students to complete the pattern.

Can you complete the pattern to show its mirror image?

Does it have symmetry?

Block symmetry 2

Materials

- a whiteboard or computer
- screen
- pattern blocks

Copy the following block arrangement on the whiteboard or screen.

Challenge the students to complete the pattern.

Can you complete the pattern to show its mirror image?

Does it have symmetry?

Flip the block

Materials

- plastic or wooden blocks
- pattern blocks
- paper
- pencils

Ask the students to work in pairs for this activity.

Can you make a pattern by flipping and tracing a block?

Encourage the students to make a tessellated pattern by repeated flipping and tracing, with one student holding the block while the other traces it.

Slide the block

Materials

- plastic or wooden blocks
- pattern blocks
- paper
- pencils

Ask the students to work in pairs for this activity.

Can you make a pattern by sliding and tracing a block?

Encourage the students to make a tessellated pattern by repeated sliding and tracing, with one student holding the block while the other traces it.

Turn the block

Materials

- plastic or wooden blocks
- pattern blocks
- paper
- pencils

Ask the students to work in pairs for this activity.

Can you make a pattern by turning and tracing a block?

Encourage the students to make a tessellated pattern by repeated flipping and turning, with one student holding the block while the other traces it.

Line them up

Materials

- a computer
- screen
- paper
- pencils

Search the term 'grid images' online and select an image of a grid.

What is the name of this shape shown on the screen?

What are the names of the different kinds of lines that form this grid?

Call for answers. If necessary, point out the different vertical, horizontal and parallel lines to the students. Ask them to record the different types of lines.

Hunting lines

Materials

- notebooks
- pencils
- paper

Revise the different kinds of lines from the previous activity.

Can you see vertical, horizontal and parallel lines inside our classroom?

Discuss the examples that the students suggest. Take the students for a walk around the school and investigate examples of vertical, horizontal and parallel lines.

Ask the students to sketch examples in their notebooks and label them.

Hunting angles

Materials

- outdoor environment
- paper
- pencils

Take the students outside the classroom to investigate angles contained in the school buildings.

An angle is where two lines meet.

Can you identify angles when you look at this building?

Tell someone where they are.

Discuss the students' responses. Ask the students to draw some of the examples of angles that they identified.

Hunting tessellations

Materials

- a computer
- screen
- outdoor environment
- paper
- pencils

Search the term 'tessellation images' online. Discuss the various images on the screen. Explain that the term 'tessellation' describes shapes fitting together without any gaps or overlaps.

Take the students outside the classroom to investigate tessellations in the school environment.

Can you identify any tessellations here?

Where are they? What do they look like?

Discuss the students' ideas. Ask them to draw some of the examples of tessellations that they identified.

Paper puzzles

Materials

- coloured paper squares or circles
- scissors

Ask each student to cut a paper square or circle into five pieces to make a puzzle.

They then exchange their puzzle with a classmate who must reconstruct it. The students can then rearrange their own puzzles to create a range of different shapes and figures.

Stick squares

Materials

- a whiteboard
- markers
- craft sticks or matchsticks

Copy this shape onto the whiteboard and display it to the students.

Ask the students to copy the shape with their sticks.

Allow time for the students to answer the following questions.

Can you take away two craft sticks so that only three squares are left?

Can you take away two craft sticks so that only two squares are left?

Can you take away eight craft sticks and leave only one square?

How many craft sticks do you need to take away to make a rectangle?

Encourage the students to make up questions of their own.

Stick triangles

Materials

- a whiteboard
- markers
- craft sticks or matchsticks

Copy this shape onto the whiteboard and display it to the students.

Ask the students to copy the shape with their sticks.

Allow time for the students to answer the following questions.

Can you take away six craft sticks so that only one triangle is left?

Can you take away three craft sticks so that only one triangle is left?

Can you take away two craft sticks so that only two triangles are left?

Can you take away three craft sticks so that only two triangles are left?

How many craft sticks do you need to take away to make a diamond?

Encourage the students to make up questions of their own.

LOCATION AND TRANSFORMATION
Year F

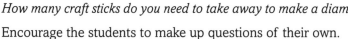

Space cities

Materials

- Lego materials or similar

Can you build an outer space city with the Lego?

When it is finished, take one of your Lego people for a walk around your city.

Can you describe the position of your Lego person as it walks around?

Discuss the students' observations.

Left and right

Materials

- paper streamers
- ribbons
- beanbags
- balls
- pairs of gloves
- paper
- pencils

Tell the students that today is a 'Left and Right Day'.

Invite the students to:

- tie a ribbon around their left hand
- throw a beanbag with their right hand and catch it with their left hand
- hop on their right foot and then on their left foot
- put on a left glove and then a right glove
- trace each other's right hand and then their left hand.

Take the opportunity to focus on left and right situations throughout the day.

Lift the barrier

Materials

- eye mask
- a wooden board or screen
- wooden blocks
- Multilink cubes or similar

Call for a pair of students to volunteer for this activity.

The students sit opposite one another with a barrier in between.

Place an eye mask on one of the students or ask them to close their eyes tightly.

The first student then builds a model with about 20 blocks or cubes, out of sight of the second student. The second student then removes the eye mask.

The first student then gives instructions to the second student, so they can copy the model on their side of the barrier. When the construction is complete the students remove the barrier and compare their models.

Were the models the same or different?

What was the same about them?

What was different about them?

Discuss the students' efforts.

OXFORD UNIVERSITY PRESS

Draw me

Materials

- paper
- pencils
- coloured pencils
- markers

Can you draw a picture of yourself:

- *on top of a box*
- *underneath a table*
- *beside a chair*
- *outside a door*
- *inside a car*
- *between two desks*
- *below a desk*
- *next to your friend*
- *at the top of the steps*
- *at the bottom of the steps?*

Discuss the students' drawings, emphasising positional language.

At the park

Materials

- paper
- pencils
- coloured pencils
- markers

Can you draw a picture of your favourite park and some of the things in it?

Ask the students to include items such as a tree, a swing, a slippery dip, a climbing frame and a bench.

Can you draw:

- *a person sitting under a tree*
- *a bird on the lowest branch of the tree*
- *a koala high up in the tree*
- *a person on the swing*
- *a person on the climbing frame*
- *a person sitting on a bench*
- *a dog beside the bench*
- *a person sliding down the slippery dip?*

Discuss the students' drawings, emphasising positional language.

The school bus

Materials

- paper
- pencils
- coloured pencils
- markers

What does your school bus look like?

Can you draw a picture of it?

Can you include:

- *the driver inside the bus*
- *a teacher standing to the left of the bus*
- *a road sign to the right of the bus*
- *a line of students near the bus*
- *a boy already on the bus*
- *a car parked beside the bus?*

Is there anything else you would like to include?

Can you describe its position?

Discuss the students' drawings, emphasising positional language.

Year 1

Tell me where

Materials

- a 4 × 4 grid
- plastic figures
- coloured counters
- coins
- Multilink cubes
- a tea towel or cover sheet

Place three or four objects on the grid and display it to the students. Allow time for the students to view the objects and then cover them.

Can you remember the position of the objects?

How would you describe their position?

Remove the cover and check the position of the objects. Discuss the students' responses and the techniques that they used.

Extension

Increase the grid size and the number of objects used.

Variation

Ask the students to draw the objects from memory before revealing them.

Tote trays

Materials

- tote trays
- tote tray cabinet
- everyday objects

Select a student to undertake this activity. Ask the other students to place their heads on their desks and close their eyes. Ask the student to hide an object inside one of the tote trays in the cabinet.

Sit up now. Where do you think the object is hiding?

Ask the other students to locate the position of the object by questioning, using positional language.

Is it on the left or the right? What is your object next to?

What is above your object? What is below your object?

Is it on the top shelf or the bottom shelf?

What is it next to? What is it in between?

Invite other students to repeat the activity.

Variation

Use books placed along the shelves of a small bookcase instead of tote trays.

Get over it

Materials

- plastic figures
- wooden blocks
- Multilink cubes
- classroom objects

Ask the students to work in small groups in order to create an obstacle course.

Can you move your figure around the course and describe its position?

In which direction does it turn?

Does it go under or over anything?

Does it go around anything?

Does it go between any blocks?

Was it able to get through any gaps?

Ask the students to pose questions to each other.

Take the opportunity during this activity to revise positional terms such as in, on, under, on top, above, below, beside, over, next to, inside, outside, beneath, between, right and left.

Extension

Read stories such as *Rosie's Walk*, *Where's Spot* and *Bear Hunt* that emphasise positional language.

Variation

Build a full-size obstacle course outside the classroom using items of furniture and sports equipment and ask the students to negotiate it.

Walk the maze

Materials
- duct tape or masking tape
- paper
- pencils

Plan and construct a large maze using tape attached to the flooring.

Can you write a series of directions to travel through the maze?

Encourage the students to use shorthand to describe the directions such as F1, B2, LT and RT.

Draw a maze

Materials
- paper
- pencil ruler

Ask the students to draw a maze. Tell them that there is a bone in the middle of it and a dog on the outside. Have them design a pathway that will lead the dog to the bone.

Year 2

Be a robot

Materials
- a pair of students

Select a pair of students for this activity.

Nominate one student as the robot and the other student as the director.

The director gives commands like 'forward two steps', 'backward one step', 'right turn', 'left turn', 'stop'.

Ask other pairs of students to join in the activity. Discuss the paths taken and any shapes made by the different movements.

OXFORD UNIVERSITY PRESS

Robot cards

Materials

- 12 robot cards

Make a set of robot cards.

Select a pair of students. Shuffle the deck of cards and ask the first student to select a card. The second student then follows the directions as given. The students then exchange roles.

Discuss the paths taken and any shapes made by the different movements.

Repeat the activity for other pairs of students.

Extension

Use a programmable robot or toy and enter a series of commands to direct to follow a path.

forward 1
forward 2
forward 3
forward 4
back 1
back 2
back 3
back 4
right turn
right turn
left turn
left turn

Bedroom map

Materials

- pencils
- paper or square dot paper

Can you imagine the position of things in your bedroom?

Can you draw a map of your bedroom to show them?

Suggest to the students that a bird's-eye view is the most appropriate way to visualise the objects and their position. Ensure that they label the positions of the door, window, bed, dressing table, desk, wardrobe and lights. Discuss and display the students' efforts.

Classroom plan

Materials

- pencil
- paper or square dot paper

Can you draw a map of your classroom and the objects in it?

Allow sufficient time for the students to observe and record their efforts.

Display and discuss their work in terms of the position of objects and people.

Classroom path

- paper
- pencils

Look across the classroom to where a friend is sitting.

Can you draw a map of how to get from your desk to your friend's desk?

Can you write some instructions about how to get there?

Discuss the students' efforts.

Secret place

Materials

- paper
- pencils

Imagine that you have hidden a secret object somewhere in your classroom.

Can you write some instructions for a friend to find it?

What is your secret object?

Variation

Allow the students to draw a map of their path, if they wish.

Spin arounds

Materials

- pairs of students
- pencils
- paper

Select five or six students and ask them to face the class.

Are you ready for this activity?

Can you spin around and complete a full turn? Which way are you now facing?

Can you complete a half turn? Which way are you facing?

What about a quarter turn? Which way are you facing?

Discuss the various orientations with the students. Ask them to represent full, half and quarter turns using pencils and paper.

MIDDLE – YEARS 3 AND 4
Measurement

USING UNITS OF MEASUREMENT
LENGTH
Year 3

How long is it?

Material

- a 1 m ruler

Close your eyes and imagine your teacher's desk. Do you think it is 1 m long, more than 1 m long, or less than 1 m long?

Allow sufficient time for the students to visualise the length of the desk.

Encourage the students to talk about their visualisation strategies. Select a student to investigate the answer. Discuss.

More than a metre long

Materials

- metre rulers or 1 m cardboard strips
- classroom objects

Which of these objects is more than 1 m long?

A ruler   A whiteboard   A student desk

Invite the students to estimate and then measure objects inside and outside the classroom that are more than 1 m long. List the students' answers on the whiteboard and discuss.

Less than a metre long

Materials

- a whiteboard
- markers
- metre rulers or 1 m cardboard strips
- classroom objects

Which of these objects is less than 1 m long?

A cupboard   A schoolbag   A whiteboard

Invite the students to estimate and then measure objects inside and outside the classroom that are less than 1 m long. List the students' answers on the whiteboard and discuss.

One metre long

Materials

- metre rulers or 1 m cardboard strips
- classroom objects
- notebooks and pencils

Can you find classroom objects that are about 1 m in length?

Invite the students to estimate and then measure objects inside and outside the classroom that are 1 m long. List the students' answers on the whiteboard and discuss.

How wide is it?

Material

- a metre rule or 1 m cardboard strip

Close your eyes and imagine the classroom doorway. Do you think it is 1 m wide, more than 1 m wide or less than 1 m wide?

Allow sufficient time for the students to visualise the width of the doorway. Encourage the students to talk about their visualisation strategies. Choose a pair of students to measure the width of the doorway. Discuss the answer.

More than a metre wide

Materials

- a whiteboard
- markers
- metre rulers or 1 m cardboard strips
- classroom objects
- notebooks and pencils

Which of these objects is more than 1 m wide?

A student desk   A cupboard   A wastepaper bin

Discuss the difference between length and width.

Invite the students to estimate and then measure objects inside and outside the classroom that are more than 1 m wide. List the students' answers on the whiteboard and discuss.

Less than a metre wide

Materials

- a whiteboard
- markers
- metre rulers or 1 m cardboard strips
- classroom objects

Which of these is less than 1 m wide?

A whiteboard   A teacher's desk   A wastepaper bin

Discuss the students' responses. Invite the students to estimate and then measure objects inside and outside the classroom that are less than 1 m wide. List the students' answers on the whiteboard and discuss.

One metre wide

Materials

- metre rulers or 1 m cardboard strips
- classroom objects
- notebooks and pencils

Can you find classroom objects that are about 1 m in width?
Invite the students to investigate the question by estimating and measuring.
Share and discuss the answers.

Extension

Ask the students to identify objects outside of their classroom that are more than 1 m, less than 1 m or about 1 m in height. Ask them to record their answers.

Our space

Materials

- metre rulers or 1 m cardboard strips
- measuring tapes

How could we measure the size of our classroom?

What do you think the dimensions would be in metres?
Ask teams of students to estimate and measure the length and width.
Record and discuss the answers.

Court dimensions

Materials

- a sports court
- metre rulers or 1 m cardboard strips
- measuring tapes

How could we measure the size of a netball or basketball court?

What do you think the dimensions would be in metres?

Ask teams of students to estimate and measure the length and width.
Record and discuss the answers.

Extension

Both types of courts may be measured and differences may be discussed. The dimensions of the courts may also be compared to those of the classroom.

Year 4

Hop it

Materials

- a whiteboard
- markers
- chalk or tape
- metre rulers or 1 m cardboard strips
- a timer

How many metres do you think you could hop in 3 seconds?

Discuss the students' responses. Select several students to assist with the activity.
Mark a starting line and choose a timekeeper. Allow the students to take turns.
Record the results in a table on the whiteboard and discuss the results.

Extension

Choose teams of students to repeat the activity.

Handspans

Materials

- student desks

What is a handspan?

Discuss the idea of a handspan, measured on an outstretched hand from the tip of the thumb to the tip of the little finger. Tell the students that this was a readily accessible unit of informal measurement used in ancient times, which is still used today.

What would be your estimate of the length of your desk in handspans?

Ask students to record their estimates and then measure their desks using handspans.

What did you discover?

Were your results different? Why?

Discuss the difficulties of measuring with informal units, as the results are different because no common standard unit is used.

Extension

Research parts of the body that were used as informal measures throughout history, such as the handspan, the palm, the foot, the digit, the cubit and the pace.

OXFORD UNIVERSITY PRESS

Less than 30 cm

Materials

- a whiteboard
- markers
- 30 cm rulers
- classroom objects
- notebooks and pencils

Can you identify five classroom objects that are less than 30 cm long?

Call for answers. Ask the students to estimate their possible lengths in centimetres.

Students can then measure the objects. Record the estimates and lengths of the objects in a table on the whiteboard. Discuss the findings with the students.

In between

Materials

- whiteboard
- markers
- metre rulers
- classroom objects
- notebooks and pencils

In pairs, can you go and find something in the classroom that is between 30 cm and 1 m in length?

Select pairs of students to demonstrate their answers.

Ask them to estimate and then measure their lengths in centimetres.

Use the whiteboard to record the estimates and lengths of the objects in a table.

Discuss the findings with the students and have them record the results.

Leafy gums

Materials

- a large collection of gum leaves
- 30 cm rulers

Show the students a gum leaf.

What would be your estimate of the length of this leaf in centimetres?

Discuss the students' responses. Select a pair of students to measure the leaf.

Discuss the result.

Follow up with group or partner work, estimating and measuring a selection of leaves that the students have collected. Ask each student to record their estimate and measurement of the leaves. Call the class together.

Which of your leaves was longest? Which of your leaves was shortest? How close were your estimates?

Discuss the students' responses.

Lengths, widths and heights

Materials

- metre rulers
- measuring tapes

Can you measure these objects in centimetres?

Write the following list on the whiteboard and ask the students to measure the items.

- The length of a desk
- The width of a desk
- The height of a desk
- The depth of a wastepaper bin
- The diameter of a wastepaper bin
- The circumference of a wastepaper bin
- The width of a door
- The thickness of a door
- The width of a dictionary
- The thickness of a dictionary

Did you discover anything that surprised you? Discuss.

Variation

Record the objects in centimetres and millimetres.

Standing long jump

Materials

- chalk or tape
- metre rulers or measuring tapes

How far can you travel when you make a standing long jump?
Select three or four students to demonstrate the process.
Measure each student's jump in centimetres.
What was the length of the longest jump?

What was the difference in centimetres between the longest jump and the shortest jump?

Extension

Students can work in pairs and complete three jumps each. They then can find the longest jump, the shortest jump and the difference between the two distances.

Size it up

Materials

- a whiteboard and markers
- a collection of Australian coins
- a 30 cm ruler

Can you name each type of Australian coin?

How many are there?

What is the smallest denomination of Australian coin?

What is the largest denomination of Australian coin?

Call for answers. List the different denominations on the whiteboard.

What would be your estimate of the diameter of the different coins?

What would be the best tool to measure the diameter?

Select students to investigate the different coin diameters.

Which coin had the smallest diameter?

Which coin had the largest diameter?

Does the largest diameter coin have the greatest value?

Does the smallest diameter coin have the least value?

List the diameters of the coins from smallest to largest.

Extension

Estimate and measure the mass of each Australian coin.

Tall and long

Materials

- a whiteboard
- markers
- a tape measure

Does the tallest person in your group have the longest arms?

Select a group of four or five students to perform the activity.

Ask the students to measure each other's heights and arm distances from shoulder to fingertips.
Record their answers on the whiteboard in decimal form and discuss them.

Name	Height (cm)	Length of arms (cm)

Geoboard perimeters

Materials

- 30 cm rulers
- 5 × 5 or 10 × 10 geoboards and coloured rubber bands or computer software

Make a shape on a geoboard and demonstrate it to the students.

How could we calculate the perimeter of this shape?

Discuss the students' answers.

Is there a way to calculate the answer without using the ruler?

Reinforce the idea of counting the side lengths.

Ask the students to create and investigate the perimeters of different shapes on geoboards or screens.

Variation

Given the perimeter, ask the students to create different shapes.

AREA
Year 3

Cover the space

Materials

- a desktop
- exercise books
- square tiles
- coloured paper circles or container lids

Some shapes fit together without any gaps. We say that these shapes 'tessellate'.

Select two students to cover the surface of a student desk, using exercise books.

Discuss the answer.

How many books do you think would be needed to cover the teacher's desk?

How many circles or lids would be needed?

How many tiles would it take?

Investigate the activity and discuss the results.

Which objects are best for covering? Why?

Develop the idea that tessellated shapes are best for covering.

Variation

Students use lids, square tiles or blocks, playing cards or pattern blocks to investigate areas.

More tessellations

Materials

- square tiles
- pattern blocks
- exercise books

How many square tiles will cover an exercise book?

Investigate the activity and discuss the results.

Extension

Students repeat the activity using different types of pattern blocks.

OXFORD UNIVERSITY PRESS

Different shapes, same area

Materials

- a whiteboard
- markers
- square tiles

Display a diagram of 6 × 4 array on the white board.

What is the area of this shape?

Can you create other shapes that have an area of 24 square units?

Encourage the students to present a range of answers. Record the solutions on the whiteboard.

Include regular and irregular shapes.

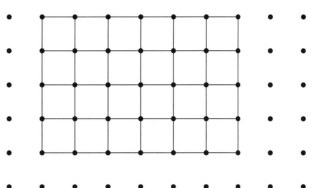

Variation

Photograph the students' solutions and make a class display.

Discuss the definition of area as the amount of space inside a 2D shape or boundary.

Geoboard areas

Materials

- 5 × 5 or 10 × 10 geoboards and coloured rubber bands or computer software

Make a shape on a geoboard and demonstrate it to the students.

How could we calculate the area of this shape?

Discuss the students' answers.

Ask the students to create and investigate the areas of different shapes on geoboards or screens.

Focus on shapes that contain right angles.

Shaded shapes

Materials

- grid paper
- pencils

Show the students a large irregular shape that you have coloured in using the squares on the grid paper or a screen image.

About how many squares do you think are coloured?

Survey the students' responses. Count the squares to calculate the result.

Encourage the students to create coloured shapes of their own and calculate their areas.

Which of your shapes had the largest area?

Which of your shapes had the smallest area?

Did you make any shapes with equal areas?

Extension

Repeat the activity using triangular or hexagonal grids.

Footprints

Materials

- pencils
- A4 paper
- transparent grid overlays

How would we find the area of a footprint?

Select two students. One student removes a shoe and places their foot on a sheet of paper. The other student traces around it.

The students then place a transparent grid overlay on top of the tracing and count the whole squares to determine the area. Add the number of squares that are more than half a unit also.

Variation

Students can trace their feet on grid paper, if transparent overlays are not available.

Thongs or flip flops may also be used.

Hands and feet

Materials

- pencils
- A4 paper
- transparent grid overlays

Is it your foot or your hand that has the larger surface area?

How could you find out?

Select two students. One student traces around the foot and hand of the other student.

The students then place a transparent grid overlay on top of the tracing and count the squares to determine the area.

What did you discover? Were they the same, or was one bigger than the other?

Extension

Repeat the activity in pairs.

Year 4

Different rectangles

Materials

- pencils
- rulers
- grid paper

How many different rectangles can you make with an area of 24 square units?

Students work alone or in pairs to investigate the question. Ask the students to label each of the rectangles constructed with four statements: e.g., six rows of four equals 24, four rows of six equals 24, $6 \times 4 = 24$ and $4 \times 6 = 24$. Include statements also for 8×3, 12×2 and 24×1 rectangles.

OXFORD UNIVERSITY PRESS

Variation

Make a class display of the students' work.

The square metre

Materials

- metre rulers
- a square metre made from newspaper and adhesive tape
- sheets of A4 paper
- coloured paper squares
- exercise books

What is a square metre?

Explain that a square metre is a square with sides that are each 1 m long.

How many exercise books will cover that square?

How could we find out?

Choose three students to investigate the activity. Ask them to estimate the number of books before counting them.

What did you find out? What was the nearest estimate that you made?

Discuss the students' responses. Repeat the activity using different objects.

What is the area?

Material

- a 1 m^2 paper square

Close your eyes and imagine your teacher's desk.

Do you think it is 1 m^2 in area, more than 1 m^2 in area or less than 1 m^2 in area?

Allow sufficient time for the students to visualise the area of the desk.

Encourage the students to talk about their visualisation strategies. Select two or three students to investigate the answer. Discuss.

More than 1 m^2

Materials

- a whiteboard
- markers
- 1 m^2 paper squares
- classroom objects
- sections of a classroom floor

Which of these objects is more than 1 m^2 in area?

A ruler   A whiteboard   A student desk

Call for answers. Select students to check one of the possibilities. Discuss the result.

Invite the students to estimate and then measure objects in the classroom that are more than 1 m^2 in area. List the students' answers on the whiteboard and discuss.

Less than 1 m²

Materials

- a whiteboard
- markers
- 1 m² paper squares
- classroom objects

Which of these objects has an area of less than 1 m²?

A doormat   A student desk   A whiteboard

Call for answers. Select students to check one of the possibilities. Discuss the result.

Invite the students to estimate and then measure objects in the classroom that are less than 1 m² in area. List the students' answers on the whiteboard and discuss.

About 1 m²

Materials

- 1 m² paper squares
- classroom objects
- notebooks and pencils

Can you find any classroom objects that are about 1 m² in area?

Call for answers. Select students to check one of the possibilities. Discuss the result.

Invite the students to investigate other possibilities by estimating and measuring. Ask them to record their answers. Share and discuss the answers.

In between

Materials

- 1 m² paper squares
- classroom objects
- notebooks and pencils

Can you find any classroom objects that are between 1 and 2 m² in area?

Invite the students to investigate the question by estimating and measuring. Ask them to record their answers. Share and discuss the answers.

Extension

Investigate areas that are between 2 and 3 m² in area.

OXFORD UNIVERSITY PRESS

Half court

Materials

- *netball court*
- *large numbers of 1 m² paper squares*

What would be your estimate of the area of half of a netball court?

What could we use to measure it?

Ensure that sufficient paper squares are available before commencing.

Count the squares and discuss the results.

Which of your estimates was closest?

How could you find out the area of a full court without measuring?

Extension

Estimate and then measure the area of a classroom floor, a walkway and a classroom wet area.

Fill the square

Materials

- measuring tape
- duct tape
- cardboard or timber strips

Construct the 1 m² frame and place it on the floor.

How many students do you think will fit inside the 1 m² frame?

Ask the class to predict their answers and write down their predictions. Select students and carry out the activity. Discuss the results.

VOLUME AND CAPACITY

Year 3

Fill the bucket

Materials

- a 2 L ice cream container
- a bucket of water
- a small plastic jug

What would be the best thing to measure how much water this bucket is holding?

A cup   An ice cream container   A margarine container

Encourage the students to estimate the number of containers required. Select two students to carry out the activity. Count the number of containers used to fill the bucket and discuss the results.

Fill the tub

Materials

- a 2 L ice cream container
- plastic or paper cups
- bucket of water

What would be the best thing to measure how much water is needed to fill this ice cream container?

A bucket   A tablespoon   A cup

Encourage the students to estimate the number of cups required. Select two students to carry out the activity. Count the number of cups and discuss the results.

Variation

Fill the container using a drink bottle, a can or a margarine or butter container.

Fill the cup

Materials

- plastic or paper cups
- Unifix or Multilink blocks
- Base 10 materials

About how many blocks will this cup hold?

Discuss the students' estimates. Count the number of blocks and discuss the results.

Filling containers

Materials

- empty butter or margarine containers
- 2 L ice cream containers
- Multilink cubes

How many Multilink cubes will fill this butter container?

Estimate and then measure the result. Discuss the students' responses.

Repeat the activity for the ice cream container and other containers.

Variation

Investigate the capacity of containers using counters, marbles or pebbles.

Compare the results.

OXFORD UNIVERSITY PRESS

Tablespoons

Materials

- plastic or paper cups
- tablespoon measures
- container of water

About how many tablespoons of water will this cup hold?

Discuss the students' estimates. Count the number of tablespoons and discuss the results.

Variation

Investigate the capacity of other cups and bowls using tablespoons of water.

On the level

Materials

- a medium-sized rock or pebble
- marbles
- a graduated kitchen jug
- water

Half fill the jug and note the level of the water.

What will happen if I place this rock in the jug?

Ask the students to predict what will happen. Place the rock in the jug and observe the result. Discuss.

How many marbles do you think it will need to raise it to the same level?

Determine how many marbles are equal in volume to a rock.

Rock it

Materials

- 3 rocks or pebbles of similar size
- a graduated kitchen jug
- water

Which of these rocks is the biggest? How could we prove it?

Select two students and ask them to arrange the rocks in the order of their size. Half fill the jug and note the level of the water. Place the first rock in the jug.

What happened? How far did the water level rise?

Repeat the activity with each of the other two rocks.

Which rock produced the largest change in water level? So which rock had the largest volume? Which rock had the smallest volume?

Discuss the students' responses.

Review the term 'volume' as the amount of space that an object occupies.

Extension

Read the story *Mr Archimedes Bath* to the students and discuss.

Measuring cylinder

Materials

- a measuring cylinder
- an empty can
- water bottle or plastic bottle
- a margarine container
- water

Display the materials to the students.

Which container do you think will have the largest capacity?

Which container do you think will have the smallest capacity?

Which containers do you think will have the same capacity?

How can we find out?

Review the term 'capacity' as the ability of a container to hold a certain quantity of liquid. Encourage the students to predict the answers to the questions.

Select two or three students and ask them to investigate by pouring.

What did you find out?

Discuss the students' responses.

Year 4

Conservation of volume

Materials

- plasticine or playdough
- a graduated kitchen jug
- water

Roll the plasticine into a small ball.

Place it into the partially filled jug and measure the rise in the water level.

Remove the ball and roll it into a sausage shape.

What do you think will be the volume of the sausage? Will the volume of the sausage be the same as the volume of the ball, or will one be bigger than the other?

Variation

Repeat the activity using 10 Multilink cubes that are unjoined and then joined.

Take your medicine

Materials

- a medicine glass
- a coffee cup
- an egg cup
- a small glass
- a large glass
- a mug
- water

Which of these empty containers will hold the smallest quantity of water?

Arrange the containers in an order that the students suggest. Select students to investigate the capacities of the containers by pouring with the medicine glass.

Were your estimates correct?

Ensure that the containers are in the correct order.

Which container has the smallest capacity?

Which container has the largest capacity?

Filling a litre

Materials

- a 1 L ice cream container or similar
- a plastic cup
- mug
- water

How many cups will it take to fill the ice cream container?

Ask the students for their estimates. Select students to investigate the answer to the question.

How many cups did you count?

If we used a mug to fill the container, would we need more mugs or fewer mugs?

Check the results and discuss the students' responses.

Extension

Repeat the activity using an egg cup and a tablespoon.

More or less

Materials

- 1 and 2 L ice cream containers
- a tote tray
- a bucket
- water

Which do you think will hold more water: the tote tray or the bucket?

Call for answers. Select two students to investigate the question. Discuss the result.

Extension

Repeat the activity using other pairs of objects.

Measuring in litres

Materials

- a whiteboard
- markers
- 1, 2 and 4 L ice cream containers
- a baking tray
- a large plastic storage container
- a tote tray
- a bucket
- water

What do you think is the capacity of these containers in litres?

Ask the students to make estimates. Record their estimates in a table on the whiteboard. Select students to measure the capacities to the nearest litre using a 1 L ice cream container or similar. Record the results and discuss.

Litre hunt

Materials

- a whiteboard
- markers

How many everyday products can you think of that come in 1 L packages?

Which products come in packages that hold more than 1 L?

Which products come in packages that hold less than 1 L?

List the results in a table under headings of less than 1 L, 1 L, 2 L, 3 L and more than 3 L.

Everyday containers

Materials

- a whiteboard
- markers
- a collection of everyday containers brought in by the students
- a measuring cylinder or graduated jug
- water

Which containers will hold more than 500 mL?

Which containers will hold less than 500 mL?

Ask the students to predict their answers.

Investigate the questions and discuss the students' responses. Record the results in a table.

Squeeze it

Materials

- oranges
- a juicer
- a knife
- 2 glasses
- a measuring cylinder or graduated jug

How much juice do you think we could squeeze from two oranges?

What would be the best way to measure it?

Ask the students to estimate the quantity of juice and identify a suitable measuring tool.

Cut the oranges and squeeze them. Pour the juice into the jug or measuring cylinder and ask a student to read the quantity.

Discuss the results with the students.

About how much juice do you think you could squeeze from 10 oranges?

Discuss the students' answers.

Extension

Compare the juiciness of individual oranges or different groups of two oranges.

Block count

Materials

- Base 10 minis

Revise the term 'volume' as the amount of space that an object occupies.

Explain that each mini block has sides of 1 cm and a volume of 1 cm^3.

How many different 3D shapes can you make with 24 minis?

Encourage the students to create block models.

What is the same about each one of your shapes?

Investigate the idea that different shapes can have the same volume.

Variation

Use Multilink cubes or similar instead of Base 10 minis.

Box designs

Materials

- Multilink cubes
- 1 or 2 cm grid paper
- scissors
- adhesive tape

Can you make a prism using 12 cubes?

Using the grid paper, can you now design an open box that will hold your prism?

Distribute the materials and allow the students to complete their designs. Discuss.

The cube family

Materials

- Multilink cubes
- Base 10 minis or other similar cubes
- isometric dot paper
- pencils

Show a single cube to the class.

If this is the smallest cube possible, what would be the size of the next two biggest cubes you could make with the blocks?

Call for answers. Select students to construct a $2 \times 2 \times 2$ cube and a $3 \times 3 \times 3$ cube.

How many cubes were used to make them?

Ask the students to sketch the cubes on isometric dot paper.

MASS

Year 3

More mass

Materials

- an equal-arm balance
- an orange
- a potato

Which is heavier: the orange or the potato?

Ask the students to predict the answer.

Select a student to heft the orange and the potato. To do this, hold one object in each hand, raise them up and down and decide which is lighter and which is heavier.

Place the orange and the potato in separate balance pans.

What happened? Which was lighter? Which was heavier?

Discuss the result.

Variation

Substitute other fruits and vegetables and investigate the results.

Equal balance

Materials

- an equal-arm balance
- an orange
- a collection of pebbles

About how many pebbles will balance an orange?

Ask the students to estimate a number. Select students to determine the result.

What would happen if we changed the size of the pebbles?

Discuss the students' responses.

Find the difference

Materials

- a whiteboard
- markers
- an equal-arm balance
- an orange
- an apple
- a collection of pebbles

Can you use pebbles to find the difference in mass between the orange and the apple?

Allow the students time to consider their answers and deduce that each can be weighed separately using the pebbles.

Was the orange heavier, or was the apple heavier, or did they have the same mass?

By how much was the difference?

Discuss the students' responses.

Record on the whiteboard:

The difference in mass between the orange and the apple was _____ pebbles.

Variation

Repeat for other pairs of objects.

MIDDLE – YEARS
3 AND 4

Package it up

Materials

- an equal-arm balance
- a small plastic bag filled with sand or rice
- pebbles
- blocks
- pasta

How many of each kind of object will balance the plastic bag?

Ask the students to estimate before measuring. Record the students' estimates in a table and then select two or three students to measure the bag using different kinds of objects.

Which of the objects were best for measuring? Why?

How many more?

Materials

- an equal-arm balance
- a calculator
- a roll of masking tape
- Multilink cubes or similar

Which will weigh more, the calculator or the roll of masking tape?

How will you know which one is heavier?

Place each object on a separate balance pan. Invite the students to use cubes to make the pans level.

Which of the objects had the larger mass?

Which of the objects had the smaller mass?

Call for answers to complete. Record on the whiteboard:

The difference in mass between the calculator and the masking tape is _____ cubes.

Repeat for other pairs of classroom objects.

Everyday measures

Materials

- a whiteboard
- markers

I weigh about 1 kg. What am I?

- A dictionary
- A letter
- An exercise book

Ask the students how they arrived at their answers.

Think of items in your kitchen or at the supermarket.

Which everyday items are packaged in amounts of 1 kg?

List the students' responses on the whiteboard. Discuss their answers.

Weighing more

Materials

- a whiteboard
- markers

I weigh more than 1 kg. What am I?

- A football
- A bag of potatoes
- An empty school backpack

Discuss the students' responses.

Which items come in packages of more than 1 kg?

Think of items in your kitchen or at the supermarket.

List the students' responses on the whiteboard and discuss.

Weighing less

Materials

- a whiteboard
- markers

I weigh less than 1 kg. What am I?

- A loaf of bread
- A bag of oranges
- A chair

Which items come in packages of less than 1 kg?

Think of items in your kitchen or at the supermarket.

List the students' responses on the whiteboard and discuss.

Kilogram hunt

Materials

- an equal-arm balance
- standard masses
- classroom objects

Which objects can we find in our classroom that weigh about 1 kg?

Invite the students to make suggestions. Select students to test the estimates by using the equal-arm balance and standard masses.

What did you discover?

Which objects weighed more than 1 kg?

Which objects weighed less than 1 kg?

Record the results.

Count how many

Materials

- kitchen scales
- marbles
- exercise books
- saucers
- cups
- oranges

How many of each of these objects will weigh 1 kg?

How much is 1 kg, measured in grams?

Select an object to be measured. Ask the class to estimate its mass.

Select students to weigh the objects.

Students record the results in a table showing each type of object, the estimate and the count.

Year 4

Half a kilogram

Materials

- kitchen scales
- blocks
- plastic cubes
- pebbles

How many of each of these objects will weigh half of 1 kg?

How much is half of 1 kg, measured in grams?

Select an object to be measured. Ask the class to estimate its mass.

Select students to weigh the objects.

Students record the results in a table showing each type of object, the estimate and the count.

Less than 500 grams

Materials

- a whiteboard
- markers
- kitchen scales
- classroom objects

Can you find any classroom objects that weigh less than 500 g?

Invite the students to make suggestions. Select students to test the estimates by using the kitchen scales.

What did you discover?

Which objects weighed less than 500 g?

List the results on the whiteboard.

In between

Materials

- a whiteboard
- markers
- kitchen scales
- classroom objects

Can you find any classroom objects that are between 500 g and 1 kg in weight?

Invite the students to make suggestions. Select students to test the estimates by using the kitchen scales.

What did you discover?

Which objects weighed between 500 g and 1 kg?

List the results on the whiteboard.

Everything in order

Materials

- a whiteboard
- markers
- kitchen scales
- a kitchen cup
- small plastic bags containing a cupful of items such as rice, pasta, flour, sand, blocks, beads, marbles or bark chips

Select two or three students and ask each to choose two bags. Compare the bags by hefting.

Which bag is heavier and which bag is lighter?

Students check their estimates using kitchen scales.

Can you order the bags from heaviest to lightest?

Were there any surprises?

Record the results in a table on the whiteboard.

100 g

Materials

- kitchen scales
- a collection of small objects such as blocks, pencils, textas, cubes

How many of each of these objects is equal to a mass of 100 g?

Select students to investigate the question. Estimate before measuring.

Discuss the students' answers.

Extension

Repeat the activity for other set amounts less than 1 kg.

Choose three

Materials

- kitchen scales
- a collection of small objects such as blocks, pencils, textas, cubes

If I choose 5 Multilink cubes, how much do you think they will weigh altogether?

Call for answers and discuss the students' estimates.

Use the kitchen scales to determine the result. Discuss the findings.

How many cubes will weigh 200 g? How many cubes will weigh 500 g?

How many cubes will weigh 1 kg?

What is the quickest way to find the answer?

Extension

Repeat the activity using a number of other small objects.

Conservation of mass

Materials

- kitchen scales
- plasticine or playdough

Select two students to roll a piece of plasticine into a small ball.

Weigh the ball using the scales.

How much did the ball weigh?

Record the result. Now ask the students to roll the ball into a sausage shape.

What do you think will be the mass of the sausage?

Will the mass of the sausage be the same as the mass of the ball?

Will the sausage weigh more or will the ball weigh more?

Ask the students to weigh the sausage. Record the result and discuss the findings.

Model masses

Materials

- a whiteboard
- markers
- kitchen scales
- Multilink cubes or similar

Show the students a model you have made using 20 Multilink cubes.

What would be the mass of this model?

Invite the students to write down their estimates. Select a student to measure the mass of the model in grams.

Was your estimate close to the result?

Add blocks to the model so that it has 50 cubes.

What would be the mass of this model?

Invite the students to again write down their estimates. Select a student to measure the mass of the model in grams.

Was your estimate close to the result?

Can you think of a quicker way to measure the mass of 100 cubes?

Record the results in a table on the whiteboard.

OXFORD UNIVERSITY PRESS

TIME
Year 3

Time of day

Materials

- a whiteboard
- markers

Write the following events on the whiteboard: first light, sunrise, mid-morning, midday, early afternoon, afternoon, sunset, twilight, night and middle of the night.

What do you think is the time of day when these events occur?

Call for answers and discuss the students' responses.

Daily routine

Materials

- pencils
- paper

At what time do you wake up in the morning of a school day?

Ask the students to record the time when they get up, eat breakfast, arrive at school, do maths work, eat lunch, go to assembly, play a sport, leave school, arrive home, watch television, do homework, play video games, eat dinner and go to bed.

Day and night

Materials

- a whiteboard
- markers
- pencils
- paper

During the day, which activity do you spend the most time doing?

Ask the students to list how much time they spend at school, playing at home, playing sport, watching TV, playing video games, eating meals, sleeping and reading. Discuss the students' responses. Use the whiteboard to graph the findings.

Follow the bouncing ball

Materials

- a stopwatch
- a tennis ball
- a basketball
- a calculator

How many times can you bounce a tennis ball in 1 minute? What about a basketball?

Which ball will give the greatest number of bounces?

Ask the students to make predictions and justify their estimates. Choose students to carry out the activity. Discuss what happened.

How many bounces would each ball make in 1 hour?

Use the calculator to help find the answer.

Join the cubes

Materials

- a stopwatch
- a collection of Multilink cubes
- a calculator

How many cubes can you join to make a tower in 2 minutes?

Select three students to perform the activity and another student as timekeeper.

Decide the average number of blocks that were joined.

Can you predict how many blocks you could join in 10 minutes? Half an hour? One hour?

Use the calculator to help find the answer.

Estimate 1 minute

Materials

- a stopwatch

Can you estimate a time period of 1 minute?

Ask the students to place their heads on their desks and close their eyes.

Tell them to raise a hand when they estimate that 1 minute has elapsed.

Who made the best estimate? How did you do it?

Introduce the technique of counting seconds by saying 'one and two and three and four...' to give a more accurate estimate.

Extension

Repeat the activity for other time periods.

How long?

Materials

- a stopwatch
- paper
- pencils

Select a group of three or four volunteers.

How long does it take to write down your spelling list?

Ask the class to record an estimate, before each student carries out the task in turn.

What did you discover? Were your estimates accurate?

Extension

Ask the students to select and investigate different activities that can be carried out in certain periods of time, e.g., reading a page from a library book.

What can you do?

Materials

- a whiteboard
- markers
- pencils
- paper

Can you think of any activity that you can complete in 1 minute, at home or school?

Discuss the students' responses. Then ask them to consider activities that could be completed in 1 second, 1 hour, 1 day, 1 week, 1 month and 1 year. Record the results in a table on the whiteboard.

Quarter hours

Materials

- a computer
- screen
- worksheets showing multiple blank analogue clock faces
- pencils
- paper

Download an interactive clock from the Internet or draw a blank analogue clock face on the whiteboard. Revise telling the time on the hour and half hour.

Today we are going to tell the time on the quarter hour.

What is the position of the hands at quarter past 11 o'clock?

Can you draw the hands on your clock face?

Do you know another way to tell the time at quarter past?

Check the answers. Ensure that students have drawn the minute hand opposite the numeral 3 and that the hour hand is one quarter of the way between 11 and 12. Accept the description of 11:15.

Suggest some other times and ask the students to represent them on their clock faces.

Repeat the activity to introduce the concept of quarter to the hour. According to common practice today, the idea of 'quarter to' can also be referred to as '45 minutes past the hour'.

Fives and tens

Materials

- a computer
- screen
- whiteboard
- markers

Download an interactive clock from the Internet or draw a blank analogue clock face on the whiteboard.

Ask the students to help you complete the markings of 5-minute intervals around the clock. Mark 0 at position 12, 5 at position 1 and so on.

Which number comes next?

How many minutes are there around the clock face?

How many minutes are there in 1 hour?

Is there a pattern to the numbers on the clock face?

Can you count by fives to 60? What about tens?

Encourage the students to count around the clock by fives and tens.

Point to a particular number on the clock face at random and have the students count around to it.

Year 4

Fives, tens and ones

Materials

- laminated blank analogue clock faces
- toy clock or wall clock
- felt pens
- paper

Distribute a laminated clock face to each pair of students.

Can you make your clock show a time of 12 minutes past 1? How do you do it?

Call for answers. Discuss the students' strategies.

Invite the students to work in pairs. Students take turns, with one student marking a time and the other student reading it.

Sequencing times

Materials

- a whiteboard
- markers

Write the following times on the whiteboard:

5:00 a.m.   7:30 p.m.   11:15 a.m.   4:51 a.m.   5:14 p.m.   12 noon

Can you place these times in their correct order?

Call for answers. Discuss the correct sequence and write it on the whiteboard.

Variation

Repeat the activity using a list of digital times.

Digital times

Materials

- a whiteboard
- markers
- pencils
- paper

On the whiteboard, write the time of 12 minutes past 9 in digital form: 9:12.

Can you write the time: 5 minutes before 9:12?

Call for answers. Give the following examples orally or write some or all questions on the whiteboard, if preferred: 3 minutes after 9:12; 12 minutes before 9:12; 13 minutes before 9:12; 18 minutes after 9:12; 1 hour after 9:12; 1 hour before 9:12; 30 minutes after 9:12; 30 minutes before 9:12.

Discuss the mental strategies used by the students to obtain their answers. List the times established in order, starting with the earliest time.

Extension

Repeat the activity for different times of the day. Encourage the students to make up questions of their own and quiz one another.

Variation

Search for an interactive clock online and use it to repeat the activity.

Clock faces

Materials

- worksheets showing multiple blank analogue clock faces
- cards displaying various digital times
- pencils
- paper

Invite two students to the front of the class. Ask each student to select a digital timecard.

Hand out the clock face worksheets.

Can you convert the digital times on the cards to analogue time on your worksheet?

Ask the students to select other cards so that the class can practice other time conversions.

Digital versus analogue

Materials

- 5 or 6 sets of matching digital and analogue timecards

Choose a number of matching pairs from one of the prepared sets. Place them face up on the desk.

Select two students to match pairs of cards.

How many matching pairs can you find here?

Display the results and discuss. Invite the class to work in groups, placing the cards face down and taking turns to match them.

a.m. and p.m.

Materials

- a whiteboard or computer
- screen

Draw two columns labelled 'Time of day' and 'Activity'.

List the times of the day at hourly intervals, starting at 6:00 a.m. and ending the next day at 5:00 a.m.

Choose a student and ask them what they might be doing at each of the times listed.

Discuss the results. Ask the class to each complete a similar table.

How many hours do you spend at school, eating meals, watching television, playing games, playing video games or sleeping?

Which activity do you spend the most time doing?

Discuss the students' findings.

Extension

Ask the students to construct a column graph showing the number of hours spent on the different activities during a 24-hour period.

School planner

Materials

- a whiteboard
- markers
- pencils
- paper
- rulers

Can you draw up a timetable of a school day?

Allocate a different day of the week to groups of students and have them create a timetable for the day. Use the whiteboard to summarise their responses and create a weekly timetable.

Will this weekly timetable be the same next week? Why? Why not?

OXFORD UNIVERSITY PRESS

MIDDLE – YEARS 3 AND 4
Shape

SHAPE
THREE-DIMENSIONAL OBJECTS
Year 3

Tracing prisms

Materials

- geometric models of square, rectangular, triangular and hexagonal prisms
- pencils
- paper

Display the model of a square prism to the students.

If we traced around the surfaces of these models, what shapes would be formed?

Select two students to carry out the activity and discuss the results.

Choose other pairs of students and repeat the activity for the other types of prisms.

What shapes can you see?

What type of shapes are they?

Display the results to the class. Develop the idea that 2D shapes may be derived from the faces of 3D objects. Discuss the students' responses.

Investigating prisms

Materials

- geometric models of square, rectangular, triangular and hexagonal prisms

Display a square prism to the students.

Tell me all you can about this model.

Allow the students sufficient time to develop their ideas.

Ask them to share their ideas with a partner, before you initiate a class discussion.

Talk about the shape of the faces, the position of faces, edges and corners and the consistency in horizontal cross-sections of prisms.

Do you know the name of this prism?

Introduce the idea that prisms are named according to the shape of their base, the base being defined as the surface that has the smallest area.

Repeat the activity using other prisms.

Building prisms

Materials

- Polydrons
- Multilink cubes
- straws
- plasticine
- pencils
- isometric dot paper

Can you imagine what a prism looks like?

Encourage the students to close their eyes and form images of prisms.

How many kinds of prisms can you make?

Encourage the students to construct the shapes with Polydrons or Multilink cubes.
They could also make skeleton shapes with straws and plasticine or lengths of spaghetti and mini-marshmallows. Label the students' models and place them on display or photograph them and create a screen display.

Extension

Ask the students to sketch models of prisms on isometric dot paper.

Prism nets

Materials

- empty grocery packets or small boxes

What would this box look like if we opened it up and laid it out flat?

Discuss the students' answers and review the responses.

Carefully pull apart a packet or box and display it to the students. Discuss.

How are the boxes put together? Why do the boxes have tabs?

Do you think the boxes are printed when they are flat, or after they have been made?

Introduce the term 'net' to describe how a 3D shape may be constructed from a flat shape.

Mystery bag 1

Materials

- a shopping bag or cardboard box
- geometric models of square, rectangular, triangular and hexagonal prisms

What does a prism feel like?

Place one of the prisms in the bag, out of the sight of students. Invite a student to reach inside the bag and carefully feel the prism.

What do you feel? Which prism do you think it is?

Variation

Place all prisms in the bag and ask the students to identify them one by one.

Prism search

Materials

- a computer
- pencils
- paper

Have you seen objects shaped like prisms in our everyday world?

Discuss the students' responses. Enter the search term 'shaped like a prism' in the computer and search for the images online. Discuss the images of various structures with the students. Ask the students to draw some of them.

Tracing pyramids

Materials

- geometric models of square, triangular and hexagonal pyramids
- pencils
- paper

Display the model of a square pyramid to the students.

If we traced around the surfaces of these models, what shapes would be formed?

Select two students to carry out the activity and discuss the results.

Choose other pairs of students and repeat the activity for the other types of pyramids.

What shapes can you see?

What type of shapes are they?

Display the results to the class and discuss the students' ideas.

Investigating pyramids

Materials

- geometric models of square, triangular and hexagonal pyramids

Show the pyramid models in turn to the students.

What can you tell me about each of these models?

Allow the students sufficient time to develop their ideas. Ask them to then share their ideas with a partner, before you initiate a class discussion.

Do you know the names of these pyramids?

Hold up and confirm the name of each pyramid in turn. Indicate the position of faces, edges and corners. Introduce the idea of naming pyramids according to the shape of their base.

Building pyramids

Materials

- Polydrons
- straws
- plasticine
- pencils
- isometric dot paper

Can you imagine what a pyramid looks like?

Encourage the students to close their eyes and form images of pyramids.

Ask the students to construct pyramids with Polydrons or make skeleton shapes with the straws and plasticine. Label the models and place them on display or photograph them and create a screen display.

Extension

Ask the students to sketch models of pyramids on isometric dot paper.

Pyramid nets

Materials

- geometric models of pyramids
- Polydrons
- pencils
- paper

Display a pyramid model to the students.

Revise the term 'net' to describe how a 3D shape may be constructed from a flat shape.

Can you imagine what would the net of a pyramid look like?

Ask the students to sketch their ideas. Choose students to create the net using Polydrons.

Mystery bag 2

Materials

- a shopping bag or cardboard box
- geometric models of square, triangular and hexagonal pyramids

What does a pyramid feel like?

Place one of the pyramids in the bag, out of the sight of students. Invite a student to reach in and carefully feel the pyramid.

What do you feel? Which pyramid do you think it is?

Variation

Place all the pyramids in the bag and ask the students to identify them one by one.

Pyramid search

Materials

- a computer
- pencils
- paper

Have you seen objects shaped like pyramids in our everyday world?

Discuss the students' responses. Enter the search term 'shaped like a pyramid' in the computer and search for images online. Discuss the images of various structures with the students. Ask the students to draw some of them.

Variation

Research the use and purpose of pyramid structures by different societies throughout history.

Mixed bag

Materials

- a shopping bag or cardboard box
- geometric models of all prisms and pyramids

Place the prisms and pyramids in the bag in full view of the students. Select a student to participate.

Can you reach in and find a prism? Can you find a pyramid? What do they feel like? How is a prism different from a pyramid?

Choose other students and repeat the activity.

Develop and demonstrate the idea that a prism has a consistent cross-section from its base upwards, whereas a pyramid has increasingly smaller cross-sections rising to its point (apex).

Investigating curved shapes

Materials

- geometric models of cylinders, cones and spheres
- pencils
- paper

Show the models in turn to the students.

What can you tell me about each of these models?

How are they different from prisms and pyramids?

Allow the students sufficient time to develop their ideas. Ask them to then share their ideas with a partner, before you initiate a class discussion. Outline the features of the sphere as having a single curved surface and cylinders and cones as having both flat and curved surfaces.

Do you know the names of these shapes?

How many curved surfaces does each shape have?

How many flat surfaces does each shape have?

Hold up and confirm the name of each shape in turn.

Ask the students to draw each of the shapes. Display the students' drawings.

Year 4

Nets from memory

Materials

- a whiteboard
- markers
- pencils
- paper

Draw a number of nets of prisms and pyramids on the whiteboard. Indicate each net in turn.

What shape would you get if this net is folded up?

Discuss the students' responses.

Extension

Ask the students to investigate combinations of six squares that will not produce a cube when folded up. Discuss this idea with the students.

Prism properties

Materials

- geometric models of a cube and square, triangular and hexagonal prisms

Display the geometric models to the students. Hold up a model and outline the definitions of face, edge and corner to the students as follows:

A face is a single flat surface on the shape.

An edge is the line segment where two faces meet.

A corner is where several edges meet.

Can we count how many faces, edges and corners these prisms have?

Create the following table on a whiteboard or screen.

Prism	Number of faces	Shape of faces	Number of edges	Number of corners
Cube				
Square prism				
Rectangular prism				
Triangular prism				
Hexagonal prism				

Encourage the students to help you complete the table.

Pyramid properties

Materials

- geometric models of square, triangular and hexagonal pyramids

Can we count how many faces, edges and corners these pyramids have?

Distribute the models to the class and select students to participate. Create the following table on a whiteboard or screen.

Pyramid	Number of faces	Shape of faces	Number of edges	Number of corners
Square pyramid				
Triangular pyramid				
Hexagonal pyramid				

Encourage the students to help you complete the table.

Visualise it

Without displaying a geometric model, ask:

How many faces are there on a cube?

How many corners are there on a cube?

How many edges are there on a cube?

Display a model of a cube or a box, if necessary, and check the answers.

Variation

Repeat this activity from time to time, covering all the prism and pyramid families.

What am I?

Play a guessing game with the students.

I have six faces and eight corners. What am I?

Repeat the game with different questions to establish the properties of different 3D solids.

Isometric dots

Materials

- isometric dot paper

Can you sketch a single cube on the dot paper?

Distribute the paper and encourage the students to represent the cube. The correct orientation of the drawing should show the edges at 30 degrees from the horizontal, matching the orientation of the dots provided. (Need a diagram here I think.)

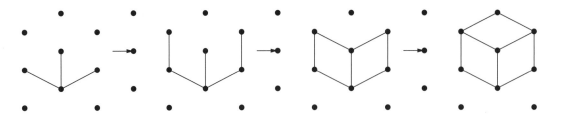

Repeat the activity using two, three and four cubes. Display the students' efforts.

Variation

Investigate the activity using online isometric drawing tools.

Letter T

Materials

- Multilink cubes
- isometric dot paper
- pencils
- paper

Can you use 5 Multilink cubes and make the letter T?

Can you sketch your shape on the dot paper?

Display the students' sketches and discuss them.

Letter L

Materials

- Multilink cubes
- isometric dot paper
- pencils
- paper

Can you use 6 Multilink cubes and make the letter L?

Can you sketch your shape on the dot paper?

Display the students' drawings.

Extension

Ask the students to create other letter shapes using blocks and sketch them on the dot paper.

Double L

Materials

- Multilink cubes
- isometric dot paper
- pencils
- paper

Can you make a double L shape with 12 cubes?

OXFORD UNIVERSITY PRESS

Students make the letter L with six cubes and sketch it. The double shape can be made by joining two single L shapes together, one on top of the other.

Variation

Extend the investigations using online isometric drawing tools.

Make and sketch

Materials

- Multilink cubes
- grid or square dot paper
- pencils

Can you build a prism using 24 cubes? How many different prisms are possible?

Display the models constructed by the students, including a $6 \times 4 \times 1$ prism, a $4 \times 3 \times 2$ prism, a $6 \times 2 \times 2$ prism, a $12 \times 2 \times 1$ prism and a $24 \times 1 \times 1$ prism.

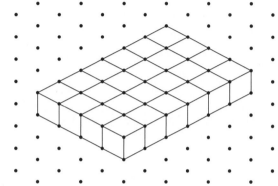

Extension

Repeat the activity using 36 cubes.

Ask the students to draw a top view and a front view of the prism they have built. Display the students' drawings.

Hidden cubes

Materials

- Multilink cubes

Select two or three students and ask them to construct $3 \times 3 \times 3$ cube. Display the cube to the class.

Do you think all the small cubes that make up this large cube can be seen?

Are there some small cubes inside the large cube?

Can you imagine how many small cubes are hidden inside the large cube?

Ask the students to make predictions. Choose a student to dismantle the cube and reveal the single hidden cube inside.

Extension

Have the students build larger models and investigate hidden cubes.

Staircases

Materials

- Multilink cubes
- isometric dot paper
- pencils
- paper

Can you build a staircase using 18 cubes, in rows of three?

Invite two or three students to carry out the activity. Ask them to arrange the 18 cubes in rows of three and join them to make the staircase.

Display the model to the class. Ask the students to draw the staircase on dot paper. Display the students' sketches and discuss them.

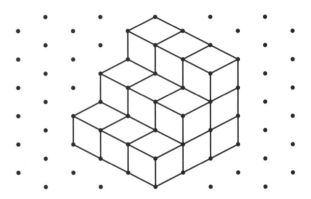

Cross-sections

Materials

- a cube
- a square pyramid and a triangular pyramid made from modelling clay
- fishing line

Select two or three students to carry out the activities.

Can you cut a cube to make a square? A triangle?

Can you cut a square pyramid to make a square? A triangle?

Can you cut a triangular pyramid to make a triangle?

Encourage the students to draw and write about what happened.

Extension

Continue to investigate cross-sections of other prisms and pyramids.

Mixed bag

Materials

- geometric models of a prism, a pyramid and a cylinder

Place the models in the bag, out of the sight of students. Invite a student to reach in and carefully select one shape. Ask the class to pose successive questions about the nature of the shape such as: Does it have corners? Is it curved? Is it pointy? How many faces can you feel?

Encourage the students to determine the nature of the shape from the descriptions given by the student handling the shape. Discourage guessing at the start of the activity.

Different views

Materials

- pencils
- paper

Close your eyes and imagine what your family car looks like.

Imagine, in turn, that you are looking at it front on, side on and from above.

Can you draw a front view of your car? Can you draw a side view of it?

What about a top (bird's-eye) view?

Discuss the students' drawings and put them on display.

Draw the carton

Materials

- an empty cardboard juice or milk carton
- pencils
- paper

Display the carton to the students.

Can you draw a front view of this carton?

Can you draw a side view of it?

What about a top (bird's-eye) view?

Discuss the students' drawings and put them on display.

TWO-DIMENSIONAL SHAPES

Year 3

Imagine this

Materials

- a whiteboard
- markers

Copy this diagram on the whiteboard.

This is a sheet of paper that has been folded in half and a piece has been cut out of it.

Can you imagine what it would look like when it is unfolded?

How would you describe the shape when it is unfolded?

Talk about the shapes predicted by the students. Ask them to draw the unfolded shape. Discuss the students' drawings.

Cut out the shape and reveal it to the students.

Did your drawings look like the cut-out shape?

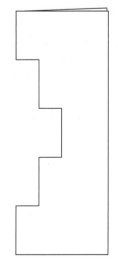

Extension

Encourage the students to make their own cut-outs with folded paper, predicting and describing the shapes that they think will be formed. Ask them to draw the predicted shapes before unfolding the paper.

3 × 3 challenge

Materials

- a whiteboard
- markers or computer drawing tools

Draw a 3 × 3 grid on the whiteboard or screen.

How many squares can you see in this grid?

Ask the students to discuss their answer with the person next to them. Allow sufficient time for their discussions.

How many of you think there are nine squares? How many of you think there are more than nine?

Discuss the students' responses. Remind them that they need to consider not just the single squares, but the single 3 × 3 square overall and the smaller 2 × 2 squares and ones that overlap.

What is your total of all the squares now?

The 3 × 3 grid has 14 squares altogether.

Geoboard triangles

Materials

- geoboards
- coloured rubber bands
- square dot paper
- pencils

Select three students to investigate this activity.

Ask the first student to make a triangle with three equal sides (equilateral).

Ask the second student to make a triangle with two equal sides (isosceles).

Ask the third student to make a triangle with no equal sides (scalene).

Display the students' work to the class and discuss.

Extension

Encourage the students to further investigate different triangles using geoboards or online learning tools for the geoboard.

Geoboard quadrilaterals

Materials

- geoboards
- coloured rubber bands
- square dot paper
- pencils

Select four students to investigate this activity.

Can you make different quadrilaterals on your geoboard?

What are their names? Did you include a parallelogram, a trapezium and a kite?

Discuss the various parallelograms with the students.

Display the students' designs or photographs.

Variation

Ask the students to use square dot paper to copy the shapes or create them with online learning tools for the geoboard.

Shapes within shapes

Materials

- a computer
- computer drawing tools
- a whiteboard
- markers
- square dot paper
- pencils

Draw a series of overlapping quadrilaterals on the whiteboard or screen.

Look at the shapes inside the shapes I have drawn.

How would you describe them?

Discuss the students' efforts with the class. Distribute the square dot paper.

Can you create some overlapping shapes of your own?

Which inside shapes did you create?

Encourage the students to talk about their designs.

Geoboard changes

Materials

- a computer
- online learning tools for the geoboard

Download an online learning tool for the geoboard. Select two or three students to undertake the activities on the screen.

Can you make a large square on the screen?

Can you change your square into a rectangle?

Can you change your rectangle into a triangle?

Can you change your triangle into a hexagon?

Discuss the students' efforts with the class.

More geoboard shapes

Materials

- a computer
- online learning tools for the geoboard

Download an online learning tool for the geoboard. Select two or three students to undertake the activities on screen.

Can you make different five-sided shapes on the screen?

What are five-sided shapes called?

Can you make different six-sided shapes on the screen?

What are six-sided shapes called?

Discuss the students' efforts with the class.

Extension

Ask the students to investigate other multi-sided polygons such heptagons and octagons online or with geoboards.

Hexagons 1

Materials

- pattern blocks
- isometric dot paper
- pencils
- rulers

Select two or three students to undertake the activity. Show them a single hexagon.

Can you use smaller pattern blocks to make a similar hexagon?

Which shapes did you use?

How is each hexagon different?

Discuss the students' efforts. Ask the class to copy the shapes made onto isometric dot paper and label them.

Extension

Encourage the students to further investigate shape combinations using online learning tools for pattern blocks.

Hexagons 2

Materials

- a set of coloured pattern blocks
- isometric dot paper
- pencils

How many large hexagons can you make with pattern blocks?

Point out a hexagon is any six-sided shape. Encourage the students to use combinations of pattern blocks to provide multiple examples of different six-sided shapes.

OXFORD UNIVERSITY PRESS

Join the dots

Materials

- isometric dot paper
- pencils

Can you use the dot paper to make:

- three different triangles?
- three different quadrilaterals?
- three different hexagons?

Observe the students' efforts and discuss the results with the class.

Variation

Repeat the activity using online isometric drawing tools.

Inside shape hunt

Materials

- a whiteboard
- markers
- pencils
- paper

Which kinds of 2D shapes can you see in our classroom?

Ask the students to list the shapes that they observe, including different types of triangles, quadrilaterals and other shapes. Discuss the students' responses.

Which type of shape is most common? Why do you think that is so?

Discuss the fact that shapes containing right angles are most common in the built environment. Summarise the students' findings using the whiteboard.

Outside shape hunt

Materials

- a whiteboard
- markers
- pencils
- paper

Which kinds of 2D shapes can you see in the natural environment?

Ask the students to list the shapes that they observe, including different types of triangles, quadrilaterals and other shapes. Discuss the students' responses.

Which type of shape is most common? Is this a different result to your classroom findings?

Summarise the students' findings using the whiteboard.

Year 4

Bathroom floors

Materials

- a computer
- online learning tools for pattern blocks
- isometric dot paper
- pencils

Show the students a tessellated pattern that you have created online.

What is special about this shape pattern?

Do you know the name used to describe shapes that fit together without any gaps?

Introduce the term to the students. Discuss the features of the pattern created.

Ask the students to create tessellated patterns on isometric dot paper for each of the different pattern block shapes.

Extension

Ask the students to design a tessellated pattern for a bathroom floor using combinations on isometric dot paper. Encourage them to display and talk about their designs.

Curvy shapes

Materials

- round counters or coins
- a computer

Display a collection of round counters to the students.

Can curved shapes tessellate?

Discuss the students' responses. Search the term 'curved tessellations' and display the images to the students.

Can you think of any examples of curved tessellations that you have seen elsewhere?

Discuss paving tiles that have curved outlines, or curved tile patterns that have circles with another curved tile filling the gaps.

What is symmetry?

Materials

- a computer
- online learning tools
- grid paper or square dot paper
- pencils
- scissors

Create a simple symmetrical design onscreen using the representations of grid or dot paper.

Discuss the design with the students.

Can you see the line of symmetry for this shape?

Can you imagine this shape folding along the line of symmetry?

Explain the idea of symmetry as mentally folding a shape in half so that each half matches exactly along a line, called the line of symmetry. The term 'reflection' may also be used.

Follow up by creating half of a symmetrical design.

Can you complete this shape so that it shows symmetry?

Ask the students to design shapes on grid or square dot paper that show symmetry.

Ask them to cut their shapes in half and investigate whether they match when they are folded. Discuss the results.

Extension

Have the students investigate symmetrical shapes using mirrors.

Block symmetry

Materials

- pattern blocks
- pencils
- paper

Ask the students to trace around the various pattern blocks.

Can you draw the lines of symmetry for each type of pattern block?

How many lines of symmetry did you find?

Variation

Repeat the activity using online representations of pattern blocks.

Draw a line of symmetry and create a mirror image pattern on each side, showing symmetry.

Geoboard symmetry

Materials

- geoboards
- coloured rubber bands

Ask the students to create some symmetrical shapes on the geoboards.

Can you make a design that has:

- *one line of symmetry*
- *two lines of symmetry*
- *more than two lines of symmetry?*

Display the students' designs and have them discuss their thinking.

Variation

Repeat the activity using online representations of pattern blocks or fold paper shapes to investigate symmetry.

Letter symmetry

Materials

- a whiteboard
- markers
- pencils
- paper

Write the following letters in large print on the whiteboard:

M   T   E   X   H   F

Can you imagine lines of symmetry for these letters?

Do any letters have more than one line of symmetry?

Is there a letter that does not have line symmetry?

Allow time for the students to investigate each example. Call for answers.

Invite the students to investigate the line symmetry of other uppercase letters.

Record the results in a table.

Letter					
Number of lines of symmetry					

Triangle symmetry

Materials

- a whiteboard
- markers
- pencils
- paper

Draw an equilateral, isosceles and scalene triangle on the whiteboard.

Do you know the names of these triangles?

How many lines of symmetry does each triangle have?

Ask the students to suggest answers and choose students to indicate their ideas at the whiteboard.

Which triangle had three lines of symmetry?

Which triangle had one line of symmetry?

Which triangle had no lines of symmetry?

Ask the students to draw the different triangles showing lines of symmetry.

Tangrams

Materials

- a computer
- online learning tool for tangrams
- cardboard
- scissors

Search the term 'tangrams' and display the images to the students.

What is the shape of the original puzzle?

What are the names of each piece of the tangram puzzle?

How many triangles are there in the puzzle?

How many quadrilaterals are there in the puzzle?

Select students to answer your questions and discuss the students' responses.

Search the NRICH site online for details of how to construct a tangram puzzle.

Encourage the students to construct their own tangram puzzles.

Tangram shapes 1

Materials

- a computer
- tangram puzzles
- paper
- pencils

Download a tangram building tool and ask two students to assist.

Using any four tangram shapes can you make: a square, a triangle, a parallelogram, a trapezium?

Ask the students to sketch the shapes that were formed.

Extension

Ask the students to use all seven tangram pieces to create a square, a triangle, a rectangle, a parallelogram and a trapezium.

Tangram shapes 2

Materials

- a computer
- tangram puzzles
- pencils
- paper

Display an image of a tangram to the students.

What shapes are made when you combine:

- *two large triangles*
- *the square and the smallest triangle*
- *the parallelogram and a small triangle?*

Encourage the students to visualise the answers to your questions before you discuss them. Ask the students to investigate their ideas with their tangram pieces.

Extension

Ask the students to sketch answers to the tangram questions.

Pattern shapes 1

Materials

- a computer
- pattern blocks or online learning tools for pattern blocks

Select two students to undertake the activity using the computer.

Can you make:

- *a larger square from four small squares*
- *a rectangle using six squares*
- *a trapezium using three triangles*
- *a parallelogram using two trapeziums*
- *a parallelogram using four rhombuses?*

Ask the students to create the above shapes using pattern blocks.

Display and discuss their designs.

Extension

Encourage the students to make larger quadrilaterals using pattern blocks.

Pattern shapes 2

Materials

- a computer
- online learning tools for pattern blocks

Select two students to undertake the activity using the computer.

What shape can you make from:

- *two or three or four or six triangles*
- *two or three or four squares*
- *three rhombuses*
- *two trapeziums?*

Ask the students to create the above shapes using pattern blocks.

Encourage further investigations. Display and discuss their designs.

LOCATION AND TRANSFORMATION
Year 3

Remember where

Materials

- a chess board
- chess pieces
- a tea towel or cover sheet
- pencils
- paper

Place three or four pieces on the grid and display it to the students. Allow time for the students to view the objects and then cover them.

Can you remember the position of the objects?

How would you describe their position?

Remove the cover and check the position of the objects. Discuss the students' responses and the techniques that they used.

Variation

Ask the students to draw the objects from memory before revealing them.

Other items may also be used.

Giving directions

Materials

- classroom plan
- pencil
- paper

Imagine that you are walking a path from the doorway of our classroom to another point in the room.

Select a student to demonstrate a path, counting paces and describing directions as forwards, turn left or turn right. For example, forward six paces, left turn, forward seven paces, right turn and so on.

Ask the students to mark a route on their classroom maps and record instructions for it.

School models

Materials

- paper
- pencils

Take the students for a walk around the school playground. Encourage them to look at the different buildings and features very carefully and memorise their position.

On return to the classroom, ask the students to sketch a map of their path, showing the position of buildings and features that they encountered. Ask them to describe their path.

Discuss the students' efforts.

Extension

Have the students create a model of the buildings based on their sketch maps.

Toy maps

Materials

- pencils
- paper or square dot paper

Close your eyes. Can you imagine the position of your favourite toy in your house?

On a floor plan of your house, can you draw a path from the front door to your toy?

Discuss and display the students' efforts.

Running messages

Materials

- pencils
- paper

Imagine that you have been asked to take messages to each classroom in your school.

What is the best path to follow?

Ask the students to draw a map of the classrooms and work out the shortest path possible.

Invite the students to discuss their proposed paths with a classmate before recording it.

Variation

If necessary, this activity can be simplified by using smaller number of classrooms.

Paper runs

Materials

- 1 cm grid paper or square dot paper
- a whiteboard
- markers
- computer drawing tools
- notebooks
- pencils

Draw an 8×8 grid on the whiteboard or create one on a screen.

Mark six crosses on grid points that represent different houses. Tell the students that they need to map out a paper run to include all of the houses.

Distribute the grid or dot paper and allow time for the students to consider their solutions.

Discuss their findings.

Did you take the most direct path?

Why would you want to do that?

Discuss the students' responses.

Year 4

Board grids

Materials

- a chessboard
- 1 cm grid paper or square dot paper
- pencils

Do you know how many squares are there on a chessboard?

In how many ways can you count them?

Which way is easiest?

Discuss counting by ones, eights or multiplying 8×8.

Ask the students to draw an 8×8 grid on their grid or square dot paper.

Use the terms 'columns' and 'rows' to describe the grid more fully.

Label the rows and columns from 1 to 8.

Can you place a cross:

- *in the second square of the second column*
- *in the fifth square of the third column*
- *in the fourth square of the fourth column*
- *in the first square of the third column*
- *in the second square of the fourth column*
- *in the fourth square of the second column?*

What shape did you make?

Extension

Encourage the students to create shapes of their own using coordinates.

Estimating north

Materials

- analogue watches
- pencils
- paper

Would you like to know a quick way to estimate the direction of north?

Take the students outside the classroom first thing in the morning.

Have the students turn towards the sun, look down and point the numeral 12 on their watches towards the position of the sun.

Ask the students to note the position of the hour hand on their watch.

Locate the position halfway between 12 and the hour hand.

Turn in that direction. You are now facing approximately north.

Extension

Display a compass to the students and explain the four main points of the compass and how to determine north.

Halfway

Materials

- a whiteboard
- markers

Draw a diagram of a compass on the whiteboard.

Indicate the main compass points of N, S, E and W and mark the mid points in between.

What do we call the directions that are halfway between: north and west, north and east, south and west and south and east?

Label the halfway positions on the whiteboard diagram and discuss.

Treasure maps

Materials

- paper or 1 cm square dot paper
- pencils

Imagine that you have hidden buried treasure somewhere in the playground of your school.

Can you draw a map of how you would get to it from our classroom?

Can you show the main features of the school on your map?

Indicate the orientation of the classroom to the students.

Ask the students to write directions using N, S, E, W and NE, SE, SW, NW and show their paths. Discuss the directions and paths the students have provided.

Ask the students to exchange their maps and directions with a partner and investigate each other's efforts.

Variation

Ask the students to include estimated distances on their maps.

Turn arounds

Can you imagine this?

In which direction is a girl now looking if she walked south, turned fully to the left, turned left again and then looked back directly over her shoulder?

Repeat the question if necessary.

Encourage the students to imagine the answer without sketching.

They can then confirm the answer using paper or square dot paper.

Extension

Have the students create and investigate similar turning problems of their own.

Where are they?

Materials

- 1 cm grid paper or square dot paper
- pencils

Here is a brainteaser about some friends.

Henry, Theo and Josie all live in the same street.

It is six blocks from Henry's house to Theo's house.

It is four blocks from Theo's house to Josie's house.

How far is it from Henry's house to Josie's house?

Encourage the students to sketch their solutions.

Is there more than one answer to this question?

Discuss the students' responses.

ANGLES
Year 3

Hunting lines

Materials

- notebooks
- pencils
- paper

Talk about different kinds of lines with the students.

Can you imagine different types of lines inside our classroom?

Discuss examples that the students suggest. Develop the discussion to include straight, curved, vertical, horizontal, diagonal, zig zag, parallel lines and perpendicular lines.

Take the students for a walk around the school and investigate examples of lines. Ask the students to sketch examples in their notebooks and label them.

Extension

Download images from the Internet and discuss different types of lines.

Spirals

Materials

- pencils
- coloured pencils or markers
- paper

Do you know what a spiral shape looks like?

Can you trace it in the air with your finger?

Encourage the students to create spirals and discuss their characteristics.

Did your curves become larger as you moved your fingers?

Distribute the materials and encourage the students to create their own spiral designs.

Make a display of the students' work.

Extension

Research the historical use of spirals by different peoples and cultures.

Line segments

Materials

- craft sticks or toothpicks
- glue
- paper or cardboard

Select two students to investigate this activity.

Ask each student to choose a capital letter of the alphabet and make it with the sticks. They then glue their letter to the paper or cardboard and display it to the class.

How many line segments are there in your letter?

Which letters have the largest number of segments?

Which letters have the smallest number of segments?

Discuss the students' responses. Ask the class members to choose a letter and further investigate the activity.

Extension

Investigate the number of line segments in numerals shown on the calculator.

Parallel lines

Materials

- a computer
- online learning tools for the geoboard
- geoboards
- coloured rubber bands

Download an online learning tool for the geoboard. Select two or three students to undertake the activities on screen.

How many different groups of parallel lines can you make on the geoboard?

Can you make a group of parallel lines:

that are of different lengths?

that are of the same length?

that are diagonal?

Ask the class to repeat the activity using geoboards and coloured rubber bands.

Display and discuss the students' efforts.

Intersections

Materials

- two sheets of plastic film
- markers
- pencils
- paper

What is formed when a pair of lines cross one another?

Draw a long line on each sheet of plastic film. Place the sheets together so that the lines intersect. Rotate the sheets to demonstrate angles as an amount of turn between two lines.

Can you imagine how many intersections are formed when a pair of parallel lines cut another pair of parallel lines?

Ask the students to imagine the lines and work in pairs to investigate the problem, drawing the lines to prove their answers.

What if three parallel lines intersected another three parallel lines?

Allow sufficient time for full consideration of the answers before discussion.

Angle wheels

Materials

- coloured paper circles
- scissors
- pencils
- paper

Demonstrate to the students how to make an angle wheel. Take two different coloured paper circles and cut a slit to the centre of each paper circle. Line up the slits and then carefully fit the circles together to make a wheel.

Using your wheel, can you make an angle with a small amount of turn?

Can you make an angle with a large amount of turn?

Can you make a right angle?

Make a display of the angles made by the students. Label the different types of angles.

Geostrip angles

Materials

- geostrips or thin cardboard strips
- split pin paper fasteners
- pencils
- paper
- rulers

Join two strips with paper fasteners. Select two students to demonstrate different angles with them.

Using the strip, can you make an angle with a small amount of turn?

Can you make an angle with a large amount of turn?

Can you make a right angle?

Ask the students to draw and label the different angles that they have created.

Geoboard angles

Materials

- geoboards
- coloured rubber bands
- computer drawing tools

How many different angles can you make on the geoboard?

Select two or three students to demonstrate different angles.

Can you make a narrow angle?

Can you make a wide angle?

Can you make a right angle?

Can you make two identical angles, one with short arms and the other with long arms?

Discuss the students' examples.

Encourage the class to use their own geoboards to copy the angles created.

Year 4

Testing angles

Materials

- A4 paper
- a whiteboard
- markers

Fold sheets of A4 paper in advance to make angles of various sizes, including a right angle. Select two or three students to undertake the activity.

Can you use these angle testers to measure angles in the classroom?

Which type of angle is most common?

On the whiteboard, make a list of right angles that the students have identified.

Variation

An angle tester can also be made by using geostrips or by placing a pipe cleaner inside a drinking straw.

Hunting angles

Materials

- A4 paper
- pencils
- angle testers (see previous activity)

Tell the students that they are going outside in order to further investigate angles.

Can you write a question about angles that is possible to investigate?

Ensure that the students consider the natural as well as the built environment.

On return to the classroom, discuss the students' questions and their findings.

OXFORD UNIVERSITY PRESS

Naming angles

Materials

- angle wheels (see activity on page 144)

Using your wheel, make an angle with a small amount of turn.

Now make an angle with a large amount of turn.

Introduce the terms 'acute' and 'obtuse' to describe these angles.

Now make some larger angles with your wheel.

Introduce the terms 'straight', 'reflex' and 'full-turn' angles to describe these angles.

Highlight the fact that angles involve the amount of turn between two lines and not the common misconception involving the distance between the arms.

Make a display of the angles made by the students. Have them label the different types of sangles.

Clock face angles

Materials

- toy clock or wall clock
- worksheets each showing multiple blank analogue clock faces

Select two students to demonstrate this activity.

Can you make a right angle on the clock face? What time is it?

Which other right angles can you make? What is the time for each angle?

Discuss the students' findings.

Ask the class to demonstrate acute, obtuse, straight, reflex and full-turn angles using their worksheets. Write the time for each angle.

Extension

Highlight the terms 'revolution and half-turn', 'full-turn' and 'quarter-turn' angles using the clock face. Revise the terms 'clockwise' and 'anticlockwise'.

UPPER – YEARS 5 AND 6
Measurement

USING UNITS OF MEASUREMENT

LENGTH

Year 5

Polygon perimeters

Materials

- rulers
- grid paper or square dot paper
- pencils/pens

Ask the students to draw four or five rectangles and calculate their perimeters.

How did you get you answer? Did you add the side lengths?

Discuss the results with the students. Develop the idea that the result can be calculated by doubling the length and doubling the width of the rectangle and then adding.

Can you draw different rectangles that each has a perimeter of 24 cm?

Discuss the different possibilities with the students.

Variation

Include examples that include decimal lengths.

Find it

Materials

- a whiteboard
- markers
- measuring tapes
- rulers
- paper
- pencils/pens

Can you find objects outside our classroom that are between 4 and 8 m in length?

Then ask them to convert the measurements to centimetres.

Encourage the students to suggest some objects before leaving the classroom.

List the students' suggestions on the whiteboard. Divide the class into small groups and investigate the activity. Have the students measure and record their findings.

On return to the classroom, ask the students to draw and write about their investigations.

OXFORD UNIVERSITY PRESS

Classroom search

Materials

- a whiteboard
- markers
- 5–10 classroom objects
- ruler or tape measure
- paper
- pens/pencils

Can you name objects in our classroom that are between 5 and 10 cm in length?

Ask the class members to nominate some objects. Collect them and display them to the class.

Display the first object.

What is your estimate of the length of this object?

Call for a show of hands. Decide on an agreed length with the class. Select two students to measure the object.

How accurate were your estimates?

Repeat the activity using other classroom objects and record the results in a table.

Object	Estimate	Measurement

Variation

Have the students collect objects that are less than 5 cm and more than 10 cm and repeat the activity. Then ask them to convert the measurements to millimetres.

Longest to shortest

Materials

- 10 small classroom objects
- a ruler
- paper
- pencils/pens

Ask the students to collect 10 small objects in the classroom and estimate and then measure their lengths in millimetres. Have them record their results in a table.

What was the longest object that you measured?

What was the shortest object that you measured?

How accurate were your estimates?

Discuss the students' findings.

Thickest to thinnest

Materials

- 10 small everyday objects
- a ruler
- paper
- pencils/pens
- calculator

Ask the students to collect 10 small objects in the classroom and estimate and then measure their thicknesses in millimetres. Have them record their results in a table.

What was the thickest object that you measured?

What was the thinnest object that you measured?

Discuss the students' findings.

Thick as a book

Materials

- 5 library books
- a ruler
- paper
- pencils/pens
- calculator

Ask the students to collect five library books and estimate and then measure their thicknesses in millimetres. Have them record their results in a table.

What was the thickest book that you measured?

What was the thinnest book that you measured?

Discuss the students' findings.

How could we calculate the thickness of a single page in these books?

Discuss the students' ideas and suggestions. Develop the idea that if the thickness is divided by the number of pages, the thickness of individual pages may be calculated. Discuss the decimal number resulting.

Body measures

Materials

- string
- scissors
- rulers

What is a cubit?

Revise the fact that the cubit is an historical measurement, comprising the distance from the tip of the index finger to the elbow, when the arm is outstretched.

Choose two students and ask them to measure each other's cubits using lengths of string, which can then be placed along the ruler, to give results in millimetres.

What did you find out?

Which person had the longer cubit?

OXFORD UNIVERSITY PRESS

Ask all class members to work in pairs, determine the lengths of the following and record their results in a table.

Body part	Estimate	Measure (mm)
Cubit		
Width of handspan		
Palm width		
Length of index finger		
Length of thumb		
Length of little finger		
Width of little finger		
Length of shoe		

Discuss the results with the class.

Did any results surprise you?

Variation

Use a tape measure instead of string, scissors and rulers.

Are you a square?

Materials

- a whiteboard
- markers
- measuring tape
- paper
- pencils/pens
- a calculator

Choose a group of three students. Ask them to measure each of their heights and the span of their outstretched arms in millimetres. Record the results on the whiteboard.

Demonstrate to the students how the ratio between height and arm span may be calculated, by dividing the height by the arm span.

If the ratio of the two quantities is approximately one, the imaginary shape formed is roughly a square. Demonstrate the results to the students.

Using paces

Materials

- student desks

What is a pace?

Discuss the idea of a pace, measured by stepping. Tell the students that this was a readily accessible unit of informal measurement used in ancient times, which is still used today.

What would be your estimate of the width of our classroom in paces?

Ask students to record their estimates. Choose three or four students to measure the width of the classroom using paces.

What did you discover?

Were your results different? Why?

Discuss the difficulties of measuring with informal units, as the results are different because no common standard unit is used.

Year 6

Trundling along

Materials

- trundle wheel
- measuring tape
- chalk

How far does a trundle wheel travel between clicks?

How could we work it out?

Discuss the students' responses. Ask them to mark the start of the wheel using a chalk line, rotate the wheel one click and mark the end point. The distance then measured is the circumference of the wheel which is exactly 1 m.

Extension

Ask the students to investigate the distance travelled by one rotation of a bike wheel.

Jogging along

Materials

- trundle wheel
- stopwatch
- paper
- pencils/pens
- stopwatch
- calculator

How long would it take for you to jog 400 m?

Use a trundle wheel to measure a 400 m track.

Ask for four or five volunteers to carry out the activity.

Record their estimated times before they start.

Start the activity and record their times using a stopwatch.

What was the average time taken to jog the distance?

What would your time be if you jogged 800 m?

Ask the same students to jog 800 m and compare their results.

Variation

Use local athletics track to determine the time taken by the students.

OXFORD UNIVERSITY PRESS

School perimeters

- trundle wheels, paper, pencils/pens

Divide the class into small groups and ask them to use trundle wheels to measure the perimeters of selected items in the school playground such as paved areas, courts, fields and buildings.

After the activities are completed, call the class together and discuss the findings.

Does measuring with the trundle wheel give the most accurate results?

Why? Why not? How do you know?

Boxed in

Materials

- a whiteboard
- markers
- grocery packets or boxes brought to school by the students
- 30 cm rulers
- tape measures
- pencils
- paper

Provide groups of students with a collection of five packets or boxes.

What are the dimensions of these packets in millimetres?

Students measure and record their answers in a table.

		Packet 1	Packet 2	Packet 3	Packet 4	Packet 5
Height (mm)	Estimate					
	Measure					
Length (mm)	Estimate					
	Measure					
Width (mm)	Estimate					
	Measure					

Which packet was the tallest? Which packet was the widest? Which packet was the longest?

Discuss the students' answers.

Classroom furnishings

Materials

- a whiteboard
- markers
- classroom furniture
- measuring tapes
- paper
- pencils/pens

Tell the students that today's activity involves estimating the height in millimetres of objects in the classroom. Select pairs of students to estimate and measure one item each.

Can you make estimates of the height of objects in the table?

Item	Estimation (mm)	Measure (mm)
Student chair		
Teacher's chair		
Cupboard		
Teacher's table		
Other		

Discuss the results and the accuracy of students' estimates.

Extension

Ask the students to investigate the length and width of the same objects.

Measure the curve

Materials

- a whiteboard
- markers
- paper
- pencils/pens
- string or wool
- rulers

Draw a curved line on the whiteboard.

How could we measure the length of this line in millimetres?

Discuss the students' suggestions. Demonstrate the technique involved using string or wool.

Ask the students to draw three of four curved lines on their paper.

Can you estimate the length of these lines in millimetres, before you measure them?

Record your estimates and results. Were your estimates accurate?

Then ask the students to convert their measurements to centimetres.

Perimeter and area

Materials

- 1 cm grid paper
- pencils
- a computer
- computer drawing tools

Do you know the difference between perimeter and area?

Talk about the definitions with the students.

Do you think that rectangles that have the same perimeter have the same area?

Discuss the student's responses.

Ask the students to draw three different rectangles that each have a perimeter of 24 cm and then calculate their areas.

What are the areas of your rectangles?

OXFORD UNIVERSITY PRESS

Were the areas of the rectangles the same as the perimeters?

Discuss the students' responses and clarify the question.

Extension

Use the online tool called 'Area Perimeter Explorer' and further investigate the relationship between area and perimeter.

Check the can

Materials

- a soft drink can
- a tape measure or piece of string
- a 30 cm ruler

Display a can to the students.

Which do you think will be larger: the height of the can or the distance around the can?

Allow sufficient time for the students to visualise their answers.

Ask the class for their predictions via a show of hands. Select two students to measure with the string and ruler and determine the length in millimetres. Discuss the students' responses.

Were you surprised? What is the word we use for the distance around an object?

Explain the term 'circumference' to the students as the distance around the boundary of a circle, or, as in common everyday use, the distance around curved objects. Discuss.

Circumferences

Materials

- string
- rulers
- tape measures

Ask the students to estimate and then measure the circumference in millimetres of a selection of everyday objects, such as a tennis ball, a football, a softball bat, a table leg, a wastepaper basket or the nearest small tree trunk. Have them record their results in a table.

Extension

Estimate and then measure the circumference of body parts, such as the fingers, thumb, wrist, the forearm or the ankle.

How far?

Materials

- a computer
- paper
- screen image of local area
- pencils/pens
- multiple copies of map

How far in kilometres is it from your school to the nearest following locations: a park; some shops; a shopping centre; a sports field; a cinema; a church; a service station; a railway station?

Work together with the students to estimate the distances and record the results. Then ask the students to convert the kilometres to metres.

Best length units

Which unit would be best to measure:

- *the width of a pencil*
- *the width of an eraser*
- *the height of a person*
- *the height of a tall building*
- *the length of a swimming pool*
- *the depth of a swimming pool*
- *the distance from Melbourne to Sydney?*

Call for answers and discuss the students' responses.

Indicate to the students that measuring in millimetres would be appropriate for the width of a pencil and an eraser; the height of people is measured in centimetres; metres are used to describe the height of buildings; the length and depth of swimming pools are measured in metres; and very large distances are measured in kilometres.

Ask the students to name some objects and identify the best length units to measure each object. Discuss the students' suggestions. Encourage them to draw and write about what they found out.

Best length devices

Materials

- trundle wheels
- metre stick
- ruler
- tape measure
- 20 m measuring tape
- paper
- pencils/pens

Display the measuring devices to the students. Ask the following questions and have the students write their answers.

Which measuring device would be best to measure:

- *the length of a bench in the playground*
- *the length of a cross-country track*
- *the width of a basketball court*
- *the width of a school corridor*
- *the height of a desk*
- *the height of a mug*
- *the distance around a person's waist?*

OXFORD UNIVERSITY PRESS

Call for answers and discuss the students' responses. Answers may vary.

Indicate to the students that the length of a bench would be best measured using a measuring tape; a trundle wheel would be best for setting out a cross-country track; a 20 m measuring tape would be best to measure the width of a basketball court and a school corridor; the height of a desk could be measured with a metre stick or a tape measure; a ruler or tape measure could measure the height of a mug; and a tape measure would be best for waist measurements.

Ask the students to name some objects and identify the best measuring devices to use in measuring each object. Discuss the students' suggestions. Encourage them to draw and write about what they found out.

Extension

Ask the students for suggestions on how to measure very large distances, such as the distance between Sydney and Brisbane. An online map, printed map or car odometer could be used.

AREA
Year 5

What is area?

Ask the students for their definition of the concept of area. Encourage them to talk about it with a classmate before calling for explanations. Area may be defined as the size of a surface, or more precisely as the space inside the boundary of a 2D shape. A common response from the students at this level is that area is length times width, which may reflect a limited understanding of the concept.

Area and perimeter

Materials

- 1 cm grid paper
- pencils/pens
- ruler
- a computer
- computer drawing tools

Do you know the difference between area and perimeter?

Talk about the definitions with the students.

Do you think that rectangles that have the same area have the same perimeter?

Discuss the student's responses. Ask the students to draw three different rectangles that each have an area of 36 cm and then calculate their areas.

What is the perimeter of your rectangles?

Were the perimeters of the rectangles the same as the areas?

Discuss the students' responses and clarify the question.

Extension

Use the online tool called 'Area Perimeter Explorer' and further investigate the relationship between area and perimeter.

Demonstrating the difference

Materials

- a whiteboard
- markers
- 4 geostrips or cardboard strips of equal lengths
- split pins
- 1 cm grid paper
- pencils/pens

Join four geostrips to make a square. Select two students to measure the area of the square with grid paper. Record the results on the whiteboard. Push the square out of shape to make a rhombus and display it to the class.

Do you think that the area of the rhombus will be greater than, less than or the same as the square?

Discuss the students' ideas. Ask the students to compare the area of the square and rhombus by tracing on grid paper to confirm that the areas are different. Encourage the students to complete the following sentence: 'Shapes that have the same perimeter do not necessarily have the same _____'.

Got it covered

Materials

- a whiteboard
- markers
- 1 cm grid paper
- scissors
- an exercise book
- 3 other books of differing sizes
- paper
- pencils/pens

Choose a pair of students and display an exercise book to the class.

What do you think is the area in square centimetres of the front cover of this book?

Call for answers. Record the students' suggestions. Ask the pair of students to measure the area of the front cover using grid paper. Discuss the results.

How accurate were your estimates?

Encourage the students to investigate the areas of other book covers and record their results. Discuss their findings.

Can you think of a more accurate way to find the area of the cover?

Discuss the students' suggestions.

Variation

Use a transparent plastic grid overlay to determine the area of the books.

OXFORD UNIVERSITY PRESS

Work it out

Materials

- an exercise book
- 3 other books of differing sizes
- rulers
- paper
- pencils/pens
- calculators

Choose the exercise book from the previous activity and display it to the class.

How can we calculate the area in square centimetres of the front cover of this book?

Call for answers. Discuss the students' suggestions. Ask a pair of students to measure the length and width of the front cover using a ruler. Have the students calculate the area in square centimetres. Calculators may be used. Encourage the students to calculate the areas of other book covers by measuring with the ruler. Ask them to record their results and discuss their findings.

How did you answers compare with the results of the previous activity when you used grid paper? Which of the two methods of finding the area was more accurate?

Scale up

Materials

- a whiteboard
- markers

Draw a 2×2 square grid on the whiteboard.

If I drew a twice-scale diagram of this grid, what would it look like?

Can you predict how many small squares it would have?

Do you think the grid would have eight squares or more than eight squares?

Discuss all answers that the students provide. Record their suggestions.

Explain to the students that the term 'twice scale' relates to the dimensions of the grid and not the total number of squares in the grid. Therefore, a twice scale diagram would be 4×4 squares in size and contain 16 small squares.

Extension

Ask the students to draw grids of their own and construct grids that are twice the scale.

Scale down

Materials

- a whiteboard
- markers

Draw a 6×4 rectangular grid on the whiteboard.

If I drew a half-scale diagram of this grid, what would it look like?

Can you predict how many small squares it would have?

Do you think the grid would have 12 squares or less than 12 squares?

Discuss all answers that the students provide. Record their suggestions.

Draw a 3 × 2 grid and ask the students to count the squares.

Were your predictions correct?

Reinforce the idea that the term 'half scale' relates to the dimensions of the grid and not the total number of squares in the grid.

Extension

Ask the students to draw grids of their own and construct grids that are half the scale.

Packets and cans

Materials

- a whiteboard
- markers
- 5 or 6 rectangular grocery packets
- a 250 mL Tetra Brik
- an aluminium can
- 1 cm grid paper
- scissors
- pencils/pens

How could we find the total surface area of each of these containers?

Discuss the students' responses. Point out that each face of the containers needs to be measured.

Divide the class into five or six groups and give each group a container to measure using grid paper.

Allow sufficient time for the students to complete the activity. Ask the students to estimate their result before using the grid paper. List their findings on the whiteboard.

Which container had the largest surface area? How accurate were your estimates?

Extension

Have the students measure the length and the width of each face of the rectangular containers and calculate their areas. Compare the results with those obtained by using grid paper.

Which of the methods you used was more accurate? Why?

Inside estimates

Materials

- rulers or tape measures
- paper
- pencils/pens
- calculators

Ask the students to identify four small rectangular surfaces inside the classroom.

Can you estimate the area of your rectangles in square centimetres?

Have the students measure the length and width of the rectangles and calculate their areas in decimal form.

What did you find out? Which rectangle was the largest? Which was the smallest?

Can you record your results in decimal form?

Draw and write about what you did.

Count it up

Materials

- a whiteboard and markers, or computer and screen
- 1 cm grid paper
- rulers
- pens/pencils

Draw the shapes shown and ask the students to copy them on a 1 cm grid paper.

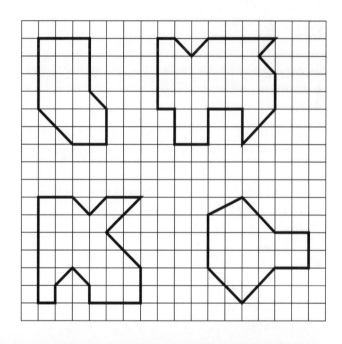

What is the area in square centimetres of each of your shapes?

Which methods did you use to count the squares that were less than a full square?

Discuss the students' results and the strategies they used.

Extension

Encourage the students to construct triangles on grid paper and repeat the activity.

Squares within squares

Materials

- a whiteboard
- markers

Copy this diagram on the whiteboard.

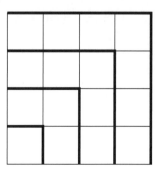

How many different size squares can you find in this large square?

Talk about your ideas with a partner.

Discuss the students' ideas. Ensure they consider 1×1, 2×2, 3×3 and 4×4 squares.

Can you find 30 squares altogether?

Draw this table on the whiteboard and ask the students to help you complete it.

Sides of square (cm)	1 × 1	2 × 2	3 × 3	4 × 4
Area (cm²)				
Difference in area (cm²)				

What patterns do you notice in the table?

Highlight the square number pattern of the areas of the shapes involved.

What is the sum of the areas of the first four squares?

What would be the difference in area of a 5 cm square?

Discuss the students' answers.

Extension

Have the students calculate and record the area of squares with sides up to 10 cm.

Year 6

On the grid

Materials

- small classroom objects such as erasers, pencils/pens, paper or notebooks, markers, pencil sharpeners, pattern blocks, 1 cm grid paper or transparent grid overlays

Choose a pair of students and display a hexagonal pattern block.

What do you think is the area of this block in square centimetres?

Call for answers and ask the class to record their estimates. Have the pair of students trace the block and count the squares.

What did you find? How accurate were your estimates?

Invite the class to repeat the activity using the other pattern blocks, recording estimates and results in a table. Discuss the students' results.

Leafy areas

Materials

- a large collection of leaves
- 1 cm grid paper or transparent grid overlays
- paper
- pencils/pens

Show the students a large leaf.

What would be your estimate of the area of this leaf in square centimetres?

Call for answers. Select a pair of students to measure the leaf. Discuss the result.

Follow up with group or partner work, estimating and measuring a selection of leaves that the students have collected. Ask each student to record their estimate and measurement of the area of each leaf. After sufficient time call the class together.

OXFORD UNIVERSITY PRESS

Which of your leaves had the largest area? Which of your leaves had the smallest area?

How close were your estimates?

Observe the accuracy of the students' estimates in order to assess their understandings.

Discuss the students' responses.

Hands and feet

Materials

- 1 cm grid paper
- pencils/pens

Is the area of your foot greater than the area of your hand, or are they the same?

How could you find out?

Ask the students to work in pairs, taking it in turns to do the tracing on grid paper.

Discuss the results after the students have counted the squares and reached their conclusions.

Extension

Investigate whether the student with the largest foot also has the largest hand.

Outside estimates

Materials

- 20 m measuring tapes
- paper
- pencils/pens
- calculators

Ask the students to identify four z × z rectangular surfaces outside the classroom. Ensure that the vertical and inclined surfaces are considered.

Can you estimate the area of your rectangles in square metres?

Have the students measure the length and width of the rectangles and calculate their areas in decimal form.

What did you find out? Which rectangle was the largest? Which was the smallest?

Draw and write about what you did.

Creating a hectare

Materials

- a whiteboard
- markers
- calculators
- a trundle wheel or 50 m measuring tape
- a large flat area
- 'witches' hat' cones

How many square metres are there in 1 hectare?

What are some possible lengths and breadths of a rectangle that is 1 hectare in area?

Encourage the children to investigate answers using their calculators.

List their suggestions and discuss them. Remind the students that a hectare does not necessarily need to be square in shape.

Find an open, flat area such as a park or a playing field and ask the students to measure its length using trundle wheels or measuring tapes. Use witches' hats to mark the corners of the rectangle.

Look at the hectare that you have created. Close your eyes and imagine its size.

Remember the amount of space that it takes up.

Fencing problem 1

Materials

- paper
- pencils
- pens
- calculators

Ask the students to imagine the following situation.

Let us pretend that you are a farmer about to fence some paddocks. You have 800 m of fencing materials. What is the largest area that can be enclosed with 800 m of fencing?

Ask the students to work in pairs and consider the possibilities.

Which shape provided the maximum possible amount of space?

Discuss the students' results.

Fencing problem 2

Materials

- a whiteboard
- markers
- paper
- pencils
- pens
- calculators

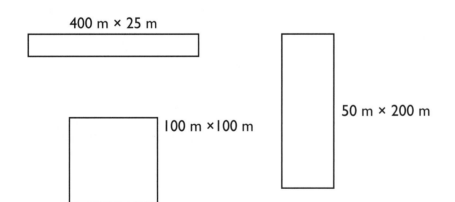

400 m × 25 m

100 m × 100 m

50 m × 200 m

OXFORD UNIVERSITY PRESS

Draw and label each of the rectangles on the whiteboard and ask the students to copy them.

Can you calculate the areas of these rectangles?

What did you find out?

Check the students' answers and confirm the fact that each of these paddocks has an area of 1 hectare. Imagine that you were a farmer again and the rectangles were your paddocks.

Which of the paddocks would be cheapest to fence? Why?

Talk about the students' ideas and suggestions.

Roll out the carpet

Materials

- 10 or 20 m measuring tape
- pencils/pens
- calculators

Can you estimate how much carpet we would need to carpet our classroom?

Ask the students to record their estimates and discuss their ideas.

How will we calculate the amount of carpet needed?

Select a pair of students to measure the length and width of the room in decimal form.

Have the students calculate the answer and compare the accuracy of their estimates.

Extension

If carpet comes in long rolls that are 4 m wide, how much carpet would need to be cut off the roll to carpet the room? Investigate the problem by drawing a diagram.

Paint the walls

Materials

- a wall
- 10 or 20 m measuring tape
- pencils/pens
- calculator

Imagine that you need to paint a wall.

What is its area?

How much paint will you need?

Choose a suitable wall and select some students to measure it in decimal form.

Have the students calculate the area of the wall and record the result.

If 1 L of paint covers about 16 m^2, how many litres of paint will be needed to give the wall two coats of paint?

If 4 L of paint costs $80 and 10 L costs $190, which would be the better buy?

Discuss the students' findings.

Extension

Calculate the amount of paint needed to paint the entire classroom.

Best area units

Which unit would be best to measure:

- *the area of a farm*
- *the floor area of a house or apartment*
- *the area of a book cover*
- *the area of a stamp or sticker*
- *the area of a small town*
- *the area of a country?*

Call for answers and discuss the students' responses. Answers may vary.

Indicate to the students that measuring in hectares would be appropriate for the area of a farm; floor areas are measured in square metres; a book cover can be measured in square centimetres; a stamp or sticker can be measured in square millimetres; a small town in hectares and a country in square kilometres.

Ask the students to name some surfaces and identify the best units to measure the area of each surface. Include vertical and inclined surfaces. Discuss the students' suggestions. Encourage them to draw and write about what they found out.

Best area devices

Materials

- trundle wheels
- metre stick
- ruler
- tape measure
- 20 m measuring tape
- a 50 m measuring tape
- paper
- pencils/pens

Display the measuring devices to the students. Ask the following questions and have the students write their answers.

Which measuring device would be best to measure the area of:

An assembly area? A football field? A stamp or sticker? A backyard swimming pool? A doormat? A book cover?

Call for answers and discuss the students' responses. Answers may vary.

Indicate to the students that the area of an assembly hall and a football field would be best measured using a 50 m measuring tape; a stamp or sticker with a ruler or tape measure; a backyard swimming pool with a 20 m tape; a doormat with a metre stick or tape measure and a book cover with a ruler.

OXFORD UNIVERSITY PRESS

VOLUME AND CAPACITY
Year 5

Filling boxes

Materials

- a collection of small open boxes
- Multilink cubes or Base 10 minis

Choose a pair of students to begin the activity. Have them select an open box.

Can you estimate how many cubes will fill your box?

Ask the class to write down their estimates.

Have the students fill the box and count the number of cubes.

Select other pairs of students and repeat the activity.

What did you find out?

How accurate were your estimates?

Discuss the results with the students.

Building layers

Materials

- 1 cm grid paper
- scissors
- Base 10 minis or similar cubes

Cut out a small rectangular grid from a sheet of grid paper and display it to the class.

Invite a pair of students to help with the activity.

How many cubes do you think will cover this grid?

If we add two more layers how many more cubes will we need to add?

How many cubes will there be altogether?

What is the name of the shape that we have made?

What is its volume?

Investigating prisms

Materials

- a computer
- paper
- pencils/pens

Search the term 'Playing with Rectangular Prisms – GeoGebra' online.

Demonstrate the construction of different prisms to the students.

Construct three different prisms in turn and ask the students to record the length, width and height of each prism in a table.

Prism	Length	Width	Height	Volume (cm³)

In how many ways could we find the volume of these prisms?

Discuss the methods of determining the volume of the prisms such as counting individual cubes or counting the layers.

How could we calculate the volume of each prism without counting the cubes?

Discuss the formula for calculating the volume of prisms. Ask the students to record the volume of each prism.

Variation

Encourage the students to undertake their own investigations of different prisms online.

Building prisms

Materials

- Multilink cubes
- Base 10 minis or similar cubes
- isometric dot paper
- pencils/pens

Can you build some prisms that have the following dimensions?

- A volume of 36 cm³, with a length of 3 cm and a height of 4 cm
- A volume of 24 cm³, with a width of 1 cm and a height of 2 cm
- A volume of 30 cm³, with a width of 2 cm and a height of 3 cm
- A volume of 40 cm³, with a length of 4 cm and a width of 2 cm

Draw some of your prisms using isometric dot paper.

Build a box

Materials

- cardboard
- 1 cm grid paper
- scissors
- adhesive tape
- Base 10 minis or similar cubes

Can you explain the difference between volume and capacity?

Revise the terms. Volume may be described as the amount of space that an object occupies, whereas capacity relates to the amount of material filling containers.

How would you make an open box that has a capacity of 18 cm³?

What would be its dimensions?

Discuss the students' suggestions. Encourage them to build the box using grid paper and adhesive tape.

Does your open box measure 3 × 3 × 2 cm or 9 × 2 × 1 cm?

What is the capacity of your box?

How could you check the capacity?

Have the students pack their boxes using Base 10 minis.

Were your predictions correct?

Variation

Glue the grid paper rectangles to cardboard to provide a more rigid result.

Measuring capacity

Materials

- calculators
- a measuring cylinder
- ruler
- dry sand
- a matchbox
- 2 or 3 small rectangular containers

Select two or three students to demonstrate this activity.

Have them first measure the length, width and height of the matchbox in millimetres and calculate the volume of the container. Record the result.

Fill the matchbox with sand and level it off. Pour the sand into the measuring cylinder and read the amount.

Were the amounts the same? Why? Why not?

Discuss the students' responses.

Repeat the activity for the remaining containers.

Variation

Rice or grain may be substituted for dry sand.

Frame it

Materials

- 12 1 m lengths of dowel
- timber
- plastic pipe or thick cardboard
- glue or masking tape
- paper
- pencils/pens

Can you construct a cube with edges that are 1 m in length?

What would be its volume?

Have a group of students build the cube.

How many students do you think can stand inside the cubic metre frame?

Ask the students to record their estimates. Count the number of students inside the frame.

How accurate were your estimates?

Discuss the results.

Imagine it

Materials

- 1 m³ framework
- plastic chairs
- paper
- pencils/pens

Show the students a single plastic chair.

How many chairs do you think will fit inside the cubic metre frame?

Ask the students to record their estimates. Choose two or three students to pack the frame with chairs. Count the number of chairs.

How accurate were your estimates?

Discuss the results.

Investigate the space

Materials

- 1 m³ framework
- paper
- pens/pencils
- classroom
- school environment

Can you imagine some objects that have volumes:
- *greater than a cubic metre*
- *less than a cubic metre*
- *about the same size as a cubic metre?*

Record your estimates in a table.

Greater than 1 m³	Less than l m³	About the same as 1 m³

Select groups of two or three students in turn to test the students' estimates, using the frame. Ask the students to draw and write about what they found out.

Cubic metres

Materials

- 1 m³ framework, paper, pens/pencils, classroom, school environment

Encourage the students to use visualisation to make estimates of the following volumes before using the cubic metre frame as a follow up.

What do you think is the approximate volume in cubic metres of:

- *a classroom*

- *a corridor*

- *a covered area*

- *an assembly hall?*

Record your estimates in a table.

Item	First estimate (m³)	Estimate with frame (m³)
Classroom		
Corridor		
Covered area		
Hall		

Discuss the results of the investigations.

Extension

Check the students' estimates by measuring the spaces with 20 or 50 m tapes and calculating the volumes.

Year 6

Everyday capacities 1

Materials

- 5–10 everyday containers
- a measuring jug
- paper
- pencils/pens

Let us investigate the capacities of some everyday containers.

Ask the students to sort the containers into three groups. The containers should hold:

- 1 L

- less than 1 L

- more than 1 L.

Have the students check the capacities of the containers by filling them with the measuring jug, recording their results in a table.

1 L	Less than 1 L	More than 1 L

Discuss the students' findings.

Everyday capacities 2

Materials

- everyday containers
- paper
- pencils/pens

Let us investigate some more capacities.

Ask the students to help you complete the following table from their previous experience and knowledge.

Less than 100 mL	Between 100 and 200 mL
Between 200 and 300 mL	Between 300 and 400 mL

Encourage the students to search for additional containers at home and record their capacities in order to complete the list.

Extension

Ask the students to list products that are sold in 250 mL, 375 mL, 1 L and 2 L containers.

Kitchen measures

Materials

- kitchen measuring cups
- kitchen measuring spoons
- a measuring cylinder or jug
- water
- paper
- pens/pencils

What is the capacity of these cups and spoons?

Ask the students to estimate the capacity of each cup and spoon, before filling them with water and pouring the contents into a measuring cylinder or jug.

Discuss the students' findings and ask them to draw and write about what they found out.

Surface areas 1

Materials

*   Multilink cubes or similar

Select three students to demonstrate this activity. Give them eight cubes each.

Can you each make a different shape using the cubes?

Allow time for the students to build their shapes.

What is the volume of these shapes?

Is it the same in each case?

Discuss the students' answers.

How could we calculate the surface areas of these shapes?

Will it be the same in each case?

Have each student calculate the surface area of their shape.

What did you find out?

Reinforce the finding that shapes having the same volume can have different surface areas.

Volume displacement

*   a Base 10 large block, ruler, a 4 L ice cream container or large saucepan, a large oven tray, a measuring cylinder or jug, water

What are the dimensions of the large Base 10 block?

If necessary, measure the cube.

What is its volume in cubic centimetres?

Call for answers and discuss the responses.

What volume of water will the block displace?

*   Fill the saucepan or ice cream container to its brim and place it in the tray.
*   Carefully place the cube in the water so that the entire block is below water and the top face is level with the surface of the water.
*   Allow the water to overflow into the tray.
*   Measure the amount of water that has been displaced by pouring the water from the tray into the measuring cylinder.

How much water overflowed into the tray?

What can you say about the relationship between 1000 cm^3 and 1 L?

Discuss the result with the students.

MASS
Year 5

What is mass?

Ask the students for their definition of the concept of mass. Encourage them to talk about it with a classmate before calling for explanations. Mass may be defined as the amount of matter in an object, or put more simply for the students, how heavy an object is.

The term 'weight' is used much more commonly in everyday life.

Mass of a bat

Materials

- a whiteboard
- markers
- a spring balance
- a cricket bat
- a softball or T-ball bat
- a hockey stick
- tennis racket
- adhesive tape

Which bat do you think will be the heaviest?

Choose a pair of students and ask them to select a bat. Ask them to estimate its mass by hefting (lifting it up and down in the hands). Record the students' estimates in a table on the whiteboard. Ask the pair to weigh the bat using the spring balance, attaching it with adhesive tape, if necessary. Record the result in the table.

Choose a new pair of students for each of the other bats and repeat the process.

Discuss the findings and list the masses of the bats in order from the heaviest to the lightest.

What did you find out? Was the result expected?

Fruity masses

Materials

- kitchen scales
- various fruits such as an apple, an orange, a banana, avocado, tomato

Select two or three students and ask them to place the pieces of fruit in order of their estimated mass. Discuss the students' choices with the class and allow them to revise the order if they wish. Measure the masses and record the results on the whiteboard.

How good were your estimates? Was the order correct?

Discuss the results with the students.

Estimating mass

Materials

- 5 or 6 everyday objects up to 500 g
- kitchen scales
- paper
- pencils/pens

How well can you estimate mass?

Choose a pair of students and have them select one of the objects. Ask them to estimate its mass by hefting. Record their estimate in a table.

Object	Estimate (g)	Measure (g)	Difference (g)

Have them measure the mass of the object, using the scales. Record the result.

Ask other pairs of students to select other objects and repeat the process, in turn. Ask the class to calculate the difference in grams between their estimates and the final measure.

How accurate were your estimates?

Extension

Repeat the activity for objects having an estimated mass of approximately 250 g.

Cupfuls

Materials

- a cup
- kitchen scales
- rice
- flour
- sugar
- breakfast cereal
- water
- paper
- pencils/pens

How much does a cupful of rice weigh?

Call for answers and ask the students to record their estimates. Choose two students to weigh the rice using the balance.

How close were your estimates?

Repeat the process. Invite other pairs of students to investigate the mass of the other materials. Ask the class to record the results in a table.

Material	Estimate (g)	Mass (g)
Rice		
Flour		
Sugar		
Breakfast cereal		
Water		

Ask the students to draw and write about what they found out.

Lunchbox masses

Materials

- a whiteboard
- markers
- a student's lunchbox
- kitchen scales
- paper
- pencils/pens
- a calculator

Ask for a volunteer to provide their lunchbox for this activity. Open the box and reveal the contents. Display them to the students.

What do you think will be the mass of this lunchbox?

Ask the students to record their estimates.

Select a student to weigh the items individually and weigh the container.

Record the results on the whiteboard. Select a pair of students to total the mass using a calculator. Record the result in decimal form.

What was the total mass?

Were your estimations accurate?

Discuss the results with the students.

Variation

Repeat the activity using several lunch boxes and calculate the average mass of a lunchbox.

100 g

Materials

- Base 10 longs and minis
- Multilink cubes
- pasta shells
- wrapped sweets
- kitchen scales
- paper
- pencils/pens

Can you estimate how many of each object is needed to equal a mass of 100 g?

Ask the students to record their estimates in a table.

Object	Estimate (g)	Measure (g)	Difference (g)

How many of each object would equal a mass of 1 kg?

Discuss the students' answers.

Extension

Ask the students to select objects of their own and repeat the activity.

Year 6

Crafty masses

Materials

- craft sticks
- kitchen scales
- paper
- pencils/pens
- calculators

How many craft sticks do you think would weigh 100 g?

Call for answers and ask the students to record their estimates.

Select two or three students to carry out the investigation.

What did you discover?

Were your estimates accurate?

How could you calculate the mass of a single craft stick?

Encourage the students to use calculators to find the answer. Discuss.

Coin masses

- a collection of Australian coins, kitchen scales, paper, pencils/pens

Choose a pair of students.

Can you estimate how many 10 c coins are needed to have a mass of approximately 50 g?

Ask the class to record their estimates. Choose a pair of students to investigate the number of 10c coins. Repeat the activity for the other types of coins.

What did you discover? How accurate were your estimates?

Base 10 masses

Materials

- Base 10 material
- kitchen scales
- paper
- pencils/pens

What do you think would be the mass of one Base 10 long?

Choose a pair of students to investigate the question using the kitchen scales. Estimate before measuring.

What did you discover?

Have the students repeat the activity for 2 longs, 4 longs, 6 longs, 8 longs and 1 flat.

Ask the class to record the results in a table.

Block	Estimate	Measure
2 longs		
4 longs		
6 longs		
8 longs		
1 flat		

Using the results in the table, how could you calculate the mass of a single mini?

Discuss the answer and methods used by the students.

Net and gross mass

Materials

- a whiteboard
- markers
- 2 different cans of food
- 2 bowls or jugs
- kitchen scales

Select a group of three or four students to investigate this activity.

Ask them to identify the mass stated on the label of each can. Record the masses in a table on the whiteboard.

	Mass on label	Mass unopened	Mass of contents	Calculated mass of container	Actual mass of container
Can 1					
Can 2					

Now ask the students to use the scales to check the mass of each can. Record the masses and discuss any differences found. Have the students open a can and use the scales to measure the mass of the contents.

Ask the class to calculate the mass of the container and have the students check the actual mass of the container with the scales.

What did you find out?

Discuss the students' observations. Introduce the terms 'net mass' to describe the mass of the contents of the container and 'gross mass' to describe the mass of the containers and its contents.

Watery masses

Materials

- a plastic or metal container
- a measuring cylinder or measuring jug
- kitchen scales
- water
- paper
- pencils/pens

Can you use the idea of net mass and gross mass to measure the mass of 1 L of water?

How would you do it?

Discuss the students' ideas and select a group of three or four students to investigate the activity.

What did you find out?

Draw and write about what you did.

Tonnes

Materials

- a computer
- paper
- pencils/pens
- a class set of calculators

Search the term 'Heaviest and lightest AFL players' online.

Reveal the lists to the students and discuss the data.

Ask each student to select a 'Dream Team' from the data and record the mass of each of the players selected.

What do you think will the combined mass of the players in your team?

What units can we use to measure the combined mass?

Have the students use calculators to total the combined masses of their teams.

Introduce the unit 'tonne', used to describe very large measures, and have the students record their results in tonnes in decimal form.

What is the average mass of players in your team?

How many players would be needed to make a mass of exactly 1 t?

Discuss the students' selections and findings.

Tare weights

Materials

- a whiteboard
- markers

Copy the table on the whiteboard.

What do you think these numbers mean?

What is the meaning of tare weight?

Ask the students to work in pairs and discuss their ideas before calling the class together for a full discussion. Explain to the students that tare weight,

WORLDWIDE SHIPPING COMPANY	
MAX. GROSS	KG 20 320
TARE	KG 2127
NET	KG 18 197

sometimes called unladen weight, is the weight of an empty container or vehicle, in this case a shipping container.

Where else might you have seen a table of numbers like the numbers on this container?

Discuss the students' suggestions. Remind them, if necessary, that some heavy trucks are required to display such information.

Extension

Ask the students to investigate the tare weights of family vehicles, printed on registration papers. Have them report their findings to the class.

Best mass units

Which unit would be best to measure the mass of:

a person; a pencil case; a calculator; a ring; a cup of rice; an orange; a whole pumpkin; a table; a pebble; a large rock; a car; a truck; an ocean liner?

Call for answers and discuss the students' responses. Answers may vary.

Indicate to the students that a person's mass is measured in kilograms; a pencil case, a calculator, a ring and an orange are measured in grams; a whole pumpkin and a table are measured in kilograms; a pebble is measured in grams; a large rock is measured in kilograms; and a car, truck and ocean liner are measured in tonnes.

Ask the students to name some objects and identify the best units to measure their mass. Discuss the students' suggestions. Encourage them to draw and write about their findings.

Best mass devices

Materials

- equal-arm balance
- kitchen scales
- spring balance
- bathroom scales

Display the measuring devices to the students. Ask the following questions orally and have the students write their answers.

OXFORD UNIVERSITY PRESS

Which measuring device would be best to measure the mass of:

a person; half a kilo of rice; an apple; a litre of water; some butter for cooking; a packed suitcase?

Call for answers and discuss the students' responses. Answers may vary.

Indicate to the students that the mass of a person can be measured using bathroom scales; half a kilo of rice, an apple, a litre of water and some butter would be measured with kitchen scales; and a spring balance would be most useful to measure a packed suitcase.

Ask the students to name some objects and identify the best measuring devices to use in measuring each object. Discuss the students' suggestions. Encourage them to draw and write about what they found out.

TIME
Year 5

Stop it

Materials

- a stopwatch

How well can you judge a time period of 10 seconds?

Choose a pair of students. The first student acts as the timer, starting the stopwatch until the second students calls time after an estimated period of 10 seconds has elapsed. Allow three attempts to encourage improvement.

Choose a second and a third pair of students and repeat the activity.

Did your estimates improve with practice? How close to 10 seconds did you get?

Extension

Repeat the activity for periods of 20 and 30 seconds.

How long?

Materials

- a stopwatch
- paper
- pencils/pens

How long does it take you to walk around the classroom and return to your starting point?

Select a group of three or four students, one student acting as timer.

Ask the class to record their estimates before each student carries out the task in turn.

What did you discover? Were your estimates accurate? About how far did you travel?

How far?

Materials
- trundle wheels
- a stopwatch
- paper
- pencils/pens

How far can you walk at a steady pace around the playground in 2 minutes?

Ask the class to record their estimate before the selected students carry out the task in turn.

Appoint a timekeeper, a student to operate the trundle wheel and another to count the clicks. Carry out the activity and discuss the results.

How accurate were your estimates?

Were the results the same? Why? Why not?

Discuss the students' ideas.

Feel the pulse

Materials
- stopwatches
- calculators
- paper
- pencils/pens

How many times do you think your pulse beats in 30 seconds?

Demonstrate the method of using the two middle fingers of the right hand to find the pulse in the left wrist. Ask the students to count and record the number of beats in 30 seconds.

Was everyone's pulse beat the same?

Discuss the results with the students.

How many times does your pulse beat in 1 minute? 10 minutes?

Use your calculator to help you find the number of beats in 1 month, 1 year and since you were born.

Birthday surprises

Materials
- a whiteboard and markers or computer and screen
- a class set of calculators

When is your birthday?

Do you know your age in years, months and days?

Call for five volunteers to help with this activity.

Draw a table like the one below and ask the students to help you complete it.

Name	Years	Months	Days

Record the name of each student and help them identify how many years, months and days have passed since their birth.

Ask the students to help you list the names of the students from the youngest to the oldest.

Can you calculate how many days old you are? Use a calculator to help.

Extension

Encourage each student in the class to calculate their age in days.

School calendar

Materials

- a whiteboard and markers or a computer and screen

What are the important events that are happening next month at our school?

Ask the students to interview people and collect information regarding excursions, meetings, debates and sporting fixtures. Prepare a calendar and ask the students to help you complete it.

Variation

Prepare a calendar listing the events of the term.

Personal timelines

Materials

- long cardboard strips
- paper
- rulers
- coloured pencils or markers
- pencils/pens

Ask the students to design a personal timeline showing important events in their life. Have them first draw a scale showing years and months before they begin. The school website may be helpful in this case also. Discuss the final products with the students before asking them to write about one of their important events.

Personal timetables

Materials

- paper
- pencils/pens

Ask the students to create their own personal timetables for the coming week. Have them mark in times, types of lessons, activities and recesses, lunchtimes and afterschool activities.

Variation

Encourage the students to create their timetable in 24-hour time.

Year 6

Clock faces

Materials

- 24-hour clock or clock face diagram
- worksheets showing combined blank analogue

Display the clock face to the students.

What is the time shown in normal time format on the clock face? What is the time in 24-hour time format?

Suggest some other time examples and ask the students to represent them on the clock faces on their worksheets.

Extension

Ask the students to research the use of 24-hour time by different organisations and why it is used.

Mixed times

Materials

- 24-hour clock or clock face diagram
- worksheets showing different blank formats, including analogue, digital and 24-hour clock faces

Ask the students to record the following times on their worksheets:

9 o'clock in the morning; 2 o'clock in the afternoon; 8 o'clock at night; half past 6 at night; quarter to 7 in the morning; 10 past 5 in the morning; 20 to 4 in the afternoon; noon; midnight.

Encourage the students to work in pairs, with one student naming a time and the other writing it in different forms.

OXFORD UNIVERSITY PRESS

Take the bus

Materials

- a whiteboard
- markers

Copy the following timetable on the whiteboard.

Ask the students to study the timetable and answer the following questions.

How long does it take the bus to travel:

- *from Dixon Rd to Virginia Ave*
- *from the bus station to Russell St*
- *from Rapp St to Smith St*
- *from Henry St to the bus station*
- *from William St to Ross St*
- *from Rapp St to Dixon Rd*
- *the completed route back to the bus station?*

Metro Bus Lines	
Departs:	
Bus station	7:30 a.m.
Rapp St	7:38 a.m.
William St	7:52 a.m.
Henry St	8:05 a.m.
Russell St	8:10 a.m.
Smith St	8:14 a.m.
Dixon Rd	8:23 a.m.
Virginia Ave	8:32 a.m.
Ross St	8:39 a.m.
Arrives bus station	8:50 a.m.

Encourage the students to create some timetables of their own and pose questions for their friends to answer.

Aboriginal and Torres Strait Islander calendars

Materials

- a computer
- screen
- paper
- pencils/pens

Search the term 'Aboriginal and Torres Strait Islander seasonal calendars' online.

Choose an appropriate calendar and display it to the students. Explain that instead of following a monthly calendar to determine the seasons, for millennia traditional Aboriginal and Torres Strait Islander people have developed a deep knowledge of the country, generally recognising six seasons in the weather cycle of each year.

The seasons are identified according to the signs in the environment that show the weather is changing. These changes involve certain plants coming into flower or fruiting, animals appearing or winds, storms and rain occurring.

Analyse and discuss the selected calendar with the students.

How many seasons can you identify? What are their names?

Which flowers or fruit identify a certain time of year?

What do people do at certain times of year?

Discuss the students' responses. Encourage them to create some questions of their own for their classmates to answer.

Time zones

Materials

- a computer
- screen
- paper
- pencils/pens

Does all of Australia have the same time?

Discuss this question with the students before researching it online.

Ask some follow-up questions:

How many major time zones is Australia divided into?

What do the initials E.S.T., C.S.T., W.S.T. and D.S.T. stand for?

What is the time difference between Sydney/Melbourne and Adelaide?

What is the time difference between Sydney/Melbourne and Perth?

If it is 11:00 a.m. in Sydney, what time is it in Perth?

If it is 11:00 a.m. in Melbourne, what time is it in Adelaide?

Extension

Ask the students to research the nature of daylight saving and which states use it. Discuss its advantages and disadvantages.

World time zones

Materials

- a computer
- screen
- paper
- pencils/pens

Search the term 'world time zones' online and select an appropriate link.

Discuss with the students how the world is divided into 24 time zones that are 1 hour apart, based on meridians of longitude.

Time is measured from the Greenwich meridian in London. Locations to the east of London are ahead of London time, while locations to the west are behind London time.

Investigate the selected map and help the students answer questions such as the following:

If it is 9:00 p.m. in Sydney, what time is it in London?

If it is 4:00 p.m. in Tokyo, what time is it in Melbourne?

If it is 10:00 a.m. in Sydney, what time is it in New York?

If it is noon in Brisbane, what time is it in Los Angeles?

If it is 1:00 p.m. in Washington, what time is it in Paris?

Discuss the students' responses and how they calculated the time differences.

Ask the students to devise and answer questions of their own.

Variation

Use the World Clock feature on a smartphone to check world times.

What is speed?

Materials

- paper
- pencils/pens
- coloured pencils
- markers

What is your definition of speed?

Ask the students to consider their definition of the concept.

Encourage them to talk about it with a classmate before calling for explanations. Discuss the students' ideas. Reinforce the fact that speed is measure of a distance travelled during a certain interval of time.

Which units are used to measure speed?

Call for answers. Confirm the fact that, on land, speed is usually measured in kilometres or miles travelled in 1 hour.

What is a speedometer?

Where do you often see one?

What are the different designs that car speedometers can have?

Discuss the students' responses and the different designs.

Ask the students to design a new look speedometer for a family car. Display their designs.

Variation

Pose questions such as if a car is travelling at 120 km/hour, how far does it travel in 1 minute?

Speed signs

Materials

- a computer
- screen
- paper
- pencils/pens

Search the term 'road signs in Australia' online.

Discuss the various signs and their use.

Why is it necessary to have speed limits?

What is the speed limit: In a residential area? Near a school? In a city? In a car park? On minor country roads? On a highway? On a motorway?

Answers may vary in some states.

Extension

Start a discussion whether speed limits should be raised or lowered in some areas. Ask the students to explain their reasons why.

Best time units

Which unit would be best to measure:

- *a school term*
- *a person's age*
- *the speed of a car*
- *a 100 m sprint*
- *the time to boil an egg*
- *the time to write your name*
- *a plane flight from Melbourne to London*
- *the time between local bus stops?*

Call for answers and discuss the students' responses. Answers may vary.

Indicate to the students that school terms are measured in weeks; a person's age is measured in years; the speed of a car is measured in kilometres per hour; boiling an egg is usually timed in minutes; writing your name would be timed in seconds; long haul flights are generally measured in hours and minutes; and the time between local bus stops is measured in minutes.

Best time devices

Materials

- a watch
- a stopwatch
- a clock
- a 24-hour clock
- a calendar

Display the measuring devices to the students. Ask the following questions and have the students write their answers.

Which measuring device would be best to measure:

- *the time taken to walk home from school*
- *the departure time for an international flight*
- *the time taken to run 100 m*
- *the time taken to read a novel*
- *the date it will be next Tuesday*
- *the time taken for a person to hop 30 m*
- *how many Saturdays there are in next month*
- *how many days are there before a person's next birthday?*

Call for answers and discuss the students' responses. Answers may vary.

Indicate to the students that the walk home from school would be measured in minutes; international flight times use 24-hour time; the time taken to run 100 m is measured in seconds and hundredths of a second; a novel would usually be read in hours; the time taken to hop 30 m would be measured in seconds; and a calendar could be consulted for the number of Saturdays in a month or the number of days until a birthday.

OXFORD UNIVERSITY PRESS

UPPER – YEARS 5 AND 6
Shape

SHAPE
THREE-DIMENSIONAL OBJECTS
Year 5

Opposite faces

Materials

- a container of dice
- paper
- pencils

Distribute the dice to the students.

Which number is opposite 2?

Which number is opposite 1?

Which number is opposite 3?

What do you notice about the pairs of numbers on opposite faces?

Allow time for the students to realise that the sum of the numbers on opposite faces is always seven.

Ask the students to draw and write about what they found out.

Cut it out

Materials

- a conical ice-cream wrapper from a Drumstick or Cornetto
- a section of cardboard tube

What does the net of a cone look like?

Ask the students to close their eyes and imagine the shape of the net.

Carefully cut and open out the ice-cream wrapper and display it.

Which 2D shapes do you see?

Were your predictions correct?

Repeat the activity using the cardboard tube.

Supermarket shapes

Materials

- a whiteboard
- markers
- a computer
- pencils/pens
- paper

Which items sold at the shops come in containers shaped like prisms?

What about containers shaped like cylinders?

Discuss the students' responses and list five examples of each item on the whiteboard.

Use the search term 'supermarket items' and discuss the shape of the containers seen.

Which items do you think will pack best into boxes for transport? Why?

Discuss the fact that shapes based on rectangular prisms pack without leaving any gaps, allowing containers to be filled to the maximum.

Ask the students to draw and label a selection of containers, emphasising the shapes they resemble. Display the students' efforts.

Curved shape properties

Materials

- a whiteboard or computer screen
- geometric models of cylinders, cones and spheres

Can we count how many faces, edges and corners there are on the cylinders, cones and spheres?

Create the following table on a whiteboard or screen.

Object	Number of faces	Shape of faces	Number of edges	Number of corners
Cylinder				
Cone				
Sphere				

Encourage the students to help you complete the table.

Which object has the greatest number of faces?

Which object has the greatest number of edges?

Which objects have no corners?

How is the sphere different from the other two objects?

OXFORD UNIVERSITY PRESS

Different cuts

Materials

- cylinders
- cones and spheres made from modelling clay
- fishing line
- pencils/pens

Select two or three students to carry out the activities. Ask the students to make predictions based on your questions.

Today we are going to cut some models with the fishing line.

Can you imagine which shapes will be formed?

Which shapes are formed when you cut a sphere?

Which shapes are formed when you cut a cylinder?

What about when you cut a cone?

Ask the students to further investigate the cross-sections formed by cutting the shapes at various angles. Encourage the students to draw and write about what happened.

The shape of the city

Materials

- a computer
- screen
- paper
- pencils
- coloured pencils

Search the term 'the shape of buildings' online.

Discuss the images viewed in terms of structures that are based on 3D shapes.

Which shapes can you identify that make up the buildings?

Are the outsides of the buildings straight or curved, or both?

Which building designs are your favourites?

Invite the students to sketch the design of their favourite building. Display their work.

Extension

Take the students on a 3D shape walk in the school environment.

Bigger cubes

Materials

- a whiteboard
- markers
- Multilink cubes

What are the dimensions of each of the cube in the cube family?

How many small cubes are there in $2 \times 2 \times 2$ cube? A $3 \times 3 \times 3$ cube?

What about a $4 \times 4 \times 4$ cube and a $5 \times 5 \times 5$ cube?

Ask the students to build the models and check the number of small cubes. Record the dimensions of the cube models and the number of blocks in each model in a table.

Cube volume	No. of small cubes
$1 \times 1 \times 1$	1
$2 \times 2 \times 2$	8
$3 \times 3 \times 3$	27
$4 \times 4 \times 4$	64
$5 \times 5 \times 5$	125

Discuss the cubic number pattern produced with the students.

Tetracubes online

Materials

- Multilink cubes
- a computer
- isometric drawing tools
- isometric dot paper
- pencils/pens

How many different arrangements can you discover using four cubes joined together?

Encourage the students to first construct tetracubes using Multilink cubes. There are eight possible arrangements.

Search the term 'Isometric Drawing Tool' online. Encourage the students to create all possible arrangements on screen.

Extension

Draw the cubes on isometric dot paper.

Painted faces

Materials

- Multilink cubes

Close your eyes and imagine this.

I have a cube that is 3 small blocks long, by 3 small blocks wide by 3 small blocks high.

How many small cubes are there in the larger cube?

Call for answers.

Now imagine that I paint all faces of the larger cube in white.

How many of the smaller cubes will have only one of their faces painted white?

Demonstrate and discuss the answer with the help of a Multilink cube model.

Odd nets

Materials

- grid paper or square dot paper
- pencils/pens
- scissors

Can you join six squares that will not fold up to make a cube?

Encourage the students to work in pairs to investigate the question.

Ask them to cut out the shapes that will not fold up to make the net of a cube.

Extension

Investigate nets that will not fold up to make various prisms and pyramids.

Twice scale

Materials

- Multilink cubes
- a whiteboard
- markers

Select two students for this activity. Ask them to make a $2 \times 2 \times 2$ cube.

How many small cubes are there in this model?

If I built a twice-scale model of this cube, what would it look like?

Can you predict how many small cubes it would have?

Accept and discuss all answers that the students provide. Record their suggestions.

Ask the two students to build a $4 \times 4 \times 4$ cube, which is twice the scale of the first cube.

How many small cubes are there in this model?

Were you surprised?

Explain to the students that the term 'twice scale' relates to the dimensions of the cube and not the total number of cubes in the model.

Half scale

Materials

- Multilink cubes

Select two students for this activity. Ask them to make a $6 \times 6 \times 6$ cube.

How many small cubes are there in this model? How did you work out your answer?

Discuss the strategies used by the students in obtaining their answers.

If I built a half-scale model of this cube, what would it look like?

Can you predict how many small cubes it would have?

Accept and discuss all answers that the students provide.

Ask the two students to build a $3 \times 3 \times 3$ cube, which is half the scale of the first cube.

How many small cubes are there in this model?

Reinforce the idea that the term 'half scale' relates to the dimensions of the cube and not the total number of cubes in the model.

Year 6

What am I?

- Play a guessing game with the students.

I am a solid shape with two flat faces and a curved face. What am I?

Include further questions such as:

I am a rectangular prism with a volume of 45 cm³. If my height is 3 cm and my length 5 cm, how wide am I?

I am a rectangular prism with a volume of 36 cm³. What are my dimensions? Is there another answer?

Encourage the students to write their own questions on cards to revise the properties of 3D solids.

Pentacubes online

Materials

- Multilink cubes
- a computer
- isometric drawing tools
- isometric dot paper
- pencils/pens

How many different arrangements can you make using five cubes joined together?

Encourage the students to first construct as many pentacubes as possible using Multilink cubes. There 29 possible arrangements of five cubes.

Search the term 'Isometric Drawing Tool' online. Allow sufficient time for the students to create a number of different pentacubes on screen. Discuss the students' efforts.

Extension

Research the Soma Cube, a 3D puzzle that can be arranged to make a cube and other block models.

Drawing pentacubes

Materials

- 3 or 4 pentacube models
- isometric dot paper
- pencils/pens
- coloured pencils

Can you draw some of your favourite pentacubes on isometric dot paper?

Discuss the students' efforts and make a display of the students' drawings.

Point of view

Materials

- geometric models of prisms, pyramids, cones, cylinders and spheres
- paper
- pencils/pens

Display the models at the front of the class.

Which of these models has:

- *a top view that looks like a circle*
- *a top view that looks like a triangle*
- *a front view that looks like a rectangle*
- *a front view that looks like a triangle?*

Hold up the models and indicate the shapes of the faces. Ask the students to draw the different views and label them.

Prism views

Materials

- geometric models of prisms
- paper
- pencils/pens

Display the geometric models of prisms to the class.

Can you draw a front view of your model?

Can you draw a side view of it?

What about a top (bird's-eye) view?

Discuss the students' drawings and put them on display.

Variation

Encourage the students to create designs on 1 cm grid paper by arranging and stacking 10 or 12 Base 10 minis. Ask the students to then sketch front, top and side views on another sheet of grid paper.

Pyramid views

Materials

- geometric models of pyramids
- paper
- pencils/pens

Display the geometric models of pyramids to the class.

Can you draw a front view of your model?

Can you draw a side view of it?

What about a top (bird's-eye) view?

Discuss the students' drawings and put them on display.

Curved views

Materials

- geometric models of cylinders, cones and spheres
- paper
- pencils/pens

Display the geometric models to the class.

Can you draw a front view of your model?

Can you draw a side view of it?

What about a top (bird's-eye) view?

Discuss the students' drawings and put them on display.

Platonic solids

Materials

- class sets of Polydrons or magnetic building sets

Which solids can you build using:

- *6 squares as faces*
- *4 triangles as faces*
- *8 triangles as faces*
- *12 pentagons as faces*
- *20 triangles as faces?*

Encourage the students to build and display the five solids.

Point out that each member of this group of solids is unique in that its faces are identical regular polygons.

Extension

Search online to find the names of individual Platonic solids that the students have constructed.

Research the term 'Archimedean solids' and encourage the students to build and name some of them.

Invisible edges

Materials

- geometric models of prisms and pyramids
- paper
- pencils/pens

Choose a model and sketch it.

Allow time for the students to complete their drawings.

How can we show the edges that are not seen in your drawings?

Introduce the idea of using dotted lines to show hidden edges in the drawings.

Place the students' drawings on display.

Block views

Materials

- Multilink cubes
- grid or square dot paper
- pencils/pens

Ask the students to construct a model using 10 or 12 blocks.

Can you draw a front view of your model?

Can you draw a side view of it?

What about a top (bird's-eye) view?

Discuss the students' drawings and put them on display.

TWO-DIMENSIONAL SHAPE

Year 5

Regular and irregular

Materials

- a whiteboard
- markers
- exercise books
- pencils/pens

Draw a variety of polygons on the whiteboard. Include an isosceles triangle, an equilateral triangle, a scalene triangle, a square, a rectangle, a parallelogram, rhombus and trapezium. Discuss the properties of each shape with the class.

Which of these polygons has sides that are all the same length and angles that are identical?

Discuss the students' responses. Identify the equilateral triangle and the square as the two examples.

Do you know the name we give to shapes that have sides that are all of the same length and all angles that are equal in size?

Introduce the term 'regular' to describe these shapes. Point out that the shapes that do not conform are called 'irregular'. Ask the students to record the definitions and examples.

Variation

Use pattern block shapes instead of whiteboard diagrams.

Regular or irregular?

Materials

- pattern blocks
- paper
- pencils/pens

Display the hexagonal pattern block to the class.

Is this a regular or irregular shape?

How do you know?

What would an irregular hexagon look like?

Can you draw some irregular hexagons?

Select some of the students' drawings and discuss the results.

Include examples that show concave features.

Extension

Develop the discussion to consider pentagons, heptagons and octagons.

Road signs

Materials

- a computer
- screen
- paper
- pencils/pens

Can you imagine the road sign nearest to our school?

What shape is it?

Does it have numbers on it? Does it have words?

Does it have numbers and words?

Search the term 'world road signs' online. Display the images and discuss the shapes, numbers and words and their purpose.

What was the most unusual road sign that you observed?

Ask the students to copy two or three of their favourite signs.

Spirals

Materials

- a computer
- screen images of spirals
- paper
- markers
- scissors
- string
- adhesive tape

Search the term 'spirals' online. Discuss some of the images displayed.

Encourage the students to create their own large spirals on paper using markers.

Ask the students to carefully cut out their spirals from the paper.

Hang the spirals from a horizontal string to create a display.

Variation

Use coloured cellophane for better contrast.

Extension

Research the history of the use of spirals in art and culture.

Stick puzzles 1

Materials

- matchsticks or craft sticks

Ask the students to copy the design shown and answer the following question.

Can you remove eight sticks so that only two squares remain?

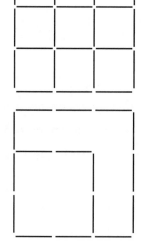

Allow sufficient time for investigation.

If necessary, reveal the following answer.

Stick puzzles 2

Materials

- matchsticks or craft sticks

Ask the students to copy the design shown and answer the following question.

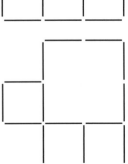

Can you remove eight sticks so that four squares remain?

Allow sufficient time for investigation.

If necessary, reveal the following answer.

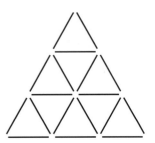

Can you discover other solutions?

Stick puzzles 3

Materials

- matchsticks or craft sticks

Ask the students to copy the design shown and answer the following question.

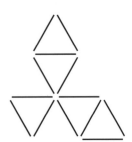

Can you remove five matches to make five small triangles?

Allow sufficient time for investigation.

If necessary, reveal the following answer.

Extension

Search 'Matchstick Puzzles' online for further challenges that the students can investigate in class time and in their free time.

Variation

Encourage the students to create their own stick puzzles.

Our flag

Materials

- a computer
- images of the Australian flag

Show the images of our national flag to the students.

What are the different shapes you can see on the Australian flag?

Which parts of the flag show symmetry?

Extension

Research the history of the Australian flag and what its different shapes represent.

Flag symmetry

Materials

- a computer
- screen images of national flags
- paper
- pencils/pens
- coloured pencils

Show a number of national flags to the students.

Can you identify which flags have symmetry?

Which flags have more than one line of symmetry?

Discuss the students' findings.

Ask the students to draw and colour their favourite symmetrical flag.

Tetrominoes

Materials

- grid paper or square dot paper
- scissors
- pencils/pens
- rulers

Have you played the game called Tetris?

How many different arrangements can you make using four squares, joined together along their sides?

Invite the students to work alone or in pairs to investigate the question.

Explain that the shapes must have common sides and not be joined at corners. When rotations and reflections are not considered, there are five possible combinations.

Were you able to discover the five possible arrangements?

Ask the students to cut out and arrange their shapes. Display the students' efforts.

Flip it

Materials

- a whiteboard
- pencils
- rulers
- coloured pencils

Copy the design shown and challenge the students to complete the pattern by visualising the shape, filling the spaces by flipping. Use coloured pencils to complete the design.

Display the students' work.

Variation

Create and complete the design using online learning tools.

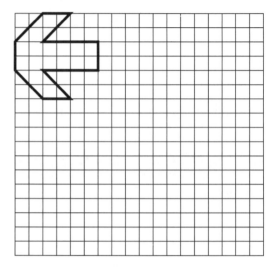

Slide it

Materials

- a whiteboard
- pencils
- rulers
- coloured pencils

Copy the design shown and challenge the students to complete the pattern by visualising the shape, filling the spaces by sliding. Use coloured pencils to complete the design.

Display the students' work.

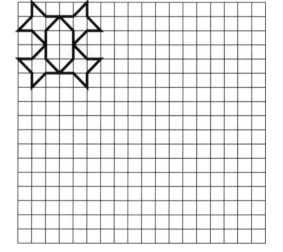

OXFORD UNIVERSITY PRESS

Turn it

Materials

- a whiteboard
- pencils
- rulers
- coloured pencils

Copy the design shown and challenge the students to complete the pattern by visualising the shape, filling the spaces by turning. Use coloured pencils to complete the design.

Display the students' work.

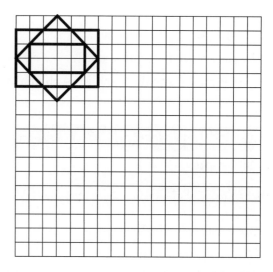

Extension

Search the term 'Shapes: Flips, Slides and Turns' online. Encourage the students to create and complete their own designs.

The correct mathematical terms of reflection, translation and rotation may be introduced at this stage.

Tetriamonds

Materials

- grid paper or square dot paper
- scissors
- pencils/pens
- rulers

How many different arrangements can you make using four triangles, joined together along their sides?

Invite the students to work alone or in pairs to investigate the question.

Explain that the shapes must have common sides and not be joined at corners. When rotations and reflections are not considered, there are three possible combinations.

Were you able to discover the three possible arrangements?

Ask the students to draw and write about what they found out.

Fold and cut

Materials

- paper squares
- scissors
- notebooks
- pencils/pens

Can you imagine this?

I have paper squares. I fold it in half to make right angle, then in half again.

I now make a straight cut across the corner near the fold.

What shape will be formed by the cut?

Distribute the materials and ask the students to carry out the activity.

What shape was formed?

Did you predict it correctly?

Ask the students to repeat the activity for three folds of paper, predicting the result before cutting.

Were you surprised by the result?

Did you predict it correctly?

Ask the students to draw and write about their findings.

Paper layers

Materials

- paper squares

Can you imagine this? (Diagrams needed)

I have a square sheet of paper. I fold it in half to make a right angle, then in half again, and then in half again. How many layers of paper are formed?

Punching holes

Materials

- paper squares
- a hole punch

Can you imagine this?

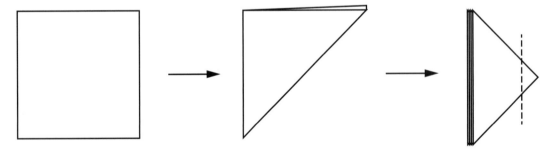

I have a square sheet of paper.

I fold it in half diagonally, then in half again and then in half again.

Now imagine I punch two holes through the paper with a hole punch.

I now open it out. How many holes will be in the sheet of paper?

Discuss the students' predictions. Select two students to carry out the activity and discuss the results.

Were your predictions accurate?

Year 6

Pentiamonds

Materials

- grid paper or square dot paper
- scissors
- pencils/pens
- rulers

How many different arrangements can you make using five triangles, joined together along their sides?

Invite the students to work alone or in pairs to investigate the question.

Explain that the shapes must have common sides and not be joined at corners. When rotations and reflections are not considered, there are four possible combinations.

Were you able to discover the four possible arrangements?

Ask the students to draw and write about what they found out.

Pancake cuts

Materials

- a whiteboard
- markers
- paper
- pencils/pens

If I cut a pancake with two intersecting cuts, how many pieces will be formed?

Call for answers. Demonstrate that there will always be four pieces with intersecting cuts, whether the cuts go through the centre or not.

How many pieces will there be if the cuts do not intersect?

Demonstrate different examples that show the three pieces formed.

What if I make three cuts, how many pieces will be formed?

Can you form four pieces? Five pieces? Seven pieces?

Explain to the students that the cuts do not have to go through the centre of the pancake or be of the same size.

Pentominoes

Materials

- grid paper or square dot paper
- scissors
- pencils/pens

How many different arrangements can you make using five squares, joined together along their sides?

Allow sufficient time for the students to investigate the possibilities.

Explain that the shapes must have common sides and not be joined at corners. When rotations and reflections are not considered, there are 12 possible combinations.

Were you able to discover the 12 possible arrangements?

Ask the students to cut out and arrange their shapes and give each of their shapes a name. Letters of the alphabet may be used. Ensure that the students store their sets for follow-up work.

Extension

Encourage the students to use pentominoes to create designs of their own, such as a house, a person, a fish or a flower.

Pentomino puzzles

Materials

- sets of pentominoes

Encourage the students to investigate pentomino puzzles in class time or free time.

Can you use three different pentominoes to make a rectangle that is 15 square units in area?

Can you use four different pentominoes to make a rectangle that is 20 square units in area?

Encourage the students to create pentomino puzzles of their own.

Pentomino symmetry

Materials

- sets of pentominoes
- paper
- pencils/pens

Encourage the students to investigate the symmetry of the pentomino pieces.

Which pentominoes have line symmetry?

Which pentominoes have rotational symmetry?

Draw and write about what you found out.

Triangle challenge

Materials

- a whiteboard
- markers or computer drawing tools

Draw the triangle as shown here on the whiteboard or screen.

How many triangles can you see in this shape?

Ask the students to discuss their answer with the person next to them. Allow sufficient time for their discussions.

How many of you think there are 13 triangles?

How many of you think there are more than 13?

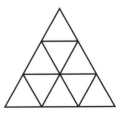

There are 13 triangles altogether: 8 small triangles, 4 triangles made from 4 small triangles and 1 large triangle.

Extension

Add another layer of triangle to the base of the shape and calculate the total number of triangles.

4 × 4 challenge

Materials

- a whiteboard
- markers or computer drawing tools

Draw a 4 × 4 grid on the whiteboard or screen.

How many squares can you see in this grid?

Ask the students to discuss their answer with the person next to them. Allow sufficient time for their discussions.

How many of you think there are 16 squares?

How many of you think there are more than 16?

Discuss the students' responses. Remind them that they need to consider not just the single squares, but the single 4 × 4 square overall and the smaller 3 × 3 and 2 × 2 squares and squares that that overlap.

What is your total of all the squares now?

The 4 × 4 grid has 30 squares altogether.

Chessboard challenge

Materials

- a whiteboard
- markers or computer drawing tools

How many squares do you think there are on a standard chessboard?

Draw an 8 × 8 grid on the whiteboard or screen.

How many squares can you see in this grid?

Ask the students to discuss their answer with the person next to them. Allow sufficient time for their discussions.

How many of you think there are 64 squares?

How many of you think there are more than 64?

Discuss the students' responses. Remind them that, as in earlier exercises, they need to systematically consider squares of all sizes from 1 × 1 up to 8 × 8. Provide help where needed.

What is your total of all the squares now?

The 4 × 4 grid has 204 squares altogether.

There are:

64 1 × 1 squares; 49 2 × 2 squares; 36 3 × 3 squares; 25 4 × 4 squares; 16 5 × 5 squares; 9 6 × 6 squares; 4 7 × 7 squares and 1 8 × 8 square.

List all the combinations on the whiteboard in a vertical form.

Can you see any pattern formed by these numbers?

Discuss the square number pattern formed.

Polygon diagonals

Materials

- geostrips or thin cardboard strips
- split pins
- a whiteboard
- markers
- a computer, screen

Choose four students to help with this activity.

Ask the students to create a triangle, a quadrilateral, a pentagon and a hexagon with the geostrips. Copy large versions of the shapes created by the students on the whiteboard.

What is a diagonal?

How many diagonals does each of these shapes have?

Invite individual students to draw in diagonals on the whiteboard.

Create the following table on a whiteboard or screen and record the results.

Shape	Number of angles	Number of sides	Number of diagonals
Triangle			
Quadrilateral			
Pentagon			
Hexagon			

Ask the students to help you complete the table.

Rotational symmetry

Materials

- 50 c coins
- paper
- pencils/pens

Ask the students to trace around a 50 c coin. Then place the coin on top of the tracing.

When you made a full turn of the coin, did it match its outline as it rotated?

How many times did it match?

Discuss the idea of a shape having rotational symmetry when its outline matches more than once in a full turn.

Extension

Research the term 'rotational symmetry' online. Discuss the ideas involved with the students. Images of automotive alloy wheels are also very useful in the conversations with the students.

OXFORD UNIVERSITY PRESS

Rotating blocks

Materials

- pattern blocks
- paper
- pencils/pens

Ask the students to trace around the various pattern blocks.

Let us investigate some more rotational symmetry by turning pattern blocks.

When you make a full turn of each block, does it match its outline as it is rotated?

How many times did it match?

Did some blocks only match once?

Did any blocks not match as they rotated?

Which blocks have rotational symmetry?

Which blocks do not have rotational symmetry?

Ask the students to draw and write about their findings.

Extension

Revise the line symmetry of pattern blocks and explain the differences between the two types of symmetry.

Rotating letters

Materials

- a whiteboard
- markers
- sheet of plastic

Write the following letters in large print on the whiteboard:

X   H   F   S   L   Z

Which of these letters have rotational symmetry?

Do some of these letters not have rotational symmetry?

Allow time for the students to investigate each example. Call for answers.

Test their responses by placing a sheet of plastic over the letter, tracing it and investigating matches by rotating it.

Invite the students to investigate the rotational symmetry of other uppercase letters.

Amazing mazes

Materials

- a computer
- grid paper
- paper
- pencils
- coloured pencils
- markers

Have you ever seen a maze? Where was it?

Did you walk through it? What happened?

Would you like to investigate some other mazes?

Discuss the students' responses.

Search the term 'mazes and labyrinths' online.

Discuss the features of the images on the screen.

What is the difference between a maze and a labyrinth?

Explain to the students that a maze is a kind of puzzle in which there a choice of paths that a person can take to get to its centre. By contrast, a labyrinth has a single path to its centre, with no junctions or alternative paths offered.

Encourage the students to draw and colour some mazes of their own. Ensure that they create some dead ends. Ask the students to write instructions on how to negotiate their mazes.

Variation

Build mazes using craft sticks, paper and glue or modelling clay.

Extension

Research the history of why mazes were built throughout history and by whom.

LOCATION AND TRANSFORMATION
Year 5

Mental pictures

Materials

- a chessboard or 8 × 8 grid
- chess pieces
- a tea towel or cover sheet
- 2 cm grid paper
- pencils/pens

Place a king, queen, bishop, knight and a castle on the grid and display it to the students. Allow time for the students to view the chess pieces and then cover them.

Can you remember the position of the objects?

Can you draw the position of each one from memory?

Distribute the grid paper and allow time for the students to sketch the position of the pieces.

Remove the cover and check the positions. Discuss the students' responses and the techniques that they used.

Can you suggest how we could more accurately describe the position of the pieces?

Discuss the students' ideas. Introduce the convention of labelling the columns using letters and the rows using numbers. Demonstrate and discuss a number of examples to the students.

Extension

Repeat the activity using increasing numbers of chess pieces.

Variation

Replace the chess pieces with everyday classroom objects.

Mental picture game

Materials

- 2 cm grid paper
- Base 10 minis or counters
- cardboard screens
- pencils/pens

Distribute a sheet of 2 cm grid paper to each pair of students. Ask them to label the columns using letters and the rows using numbers.

Ask the students to place a screen on their desk between each other. The first student in the pair then places 10 counters on the grid, out of sight of the other student.

Using coordinates, the other student is given 10 attempts to guess the location of the counters. Each successful attempt scores one point.

The students exchange roles and the one with the highest score is the winner.

Classroom grid plan

Materials

- a whiteboard
- ruler
- markers
- paper
- pencils/pens

Draw a 10 × 10 grid on the whiteboard. Label the columns using the letters from A to J and the rows from numbers 1 to 10.

Invite the students to help you make a simple plan of the classroom, showing different objects and major items of equipment. Ask the students to write down the grid references for the objects shown on the grid.

What are the coordinates of your desk?

Discuss the students' responses. Choose different coordinates and ask the students to identify the corresponding objects on their plans.

Extension

Search the term 'Battleship Game' online and have the students practise using coordinates.

Take your seat

Materials

- a whiteboard
- ruler
- markers

Draw a 12 × 8 grid on the whiteboard. Ask the students to imagine that you have created a cinema seating plan.

How can we identify the location of each seat?

Discuss the students' responses.

Label the columns using the letters from A to H and the rows from numbers 1 to 12.

Ask the students to imagine that they are booking seats online for a movie.

Invite each student in the class to make a reservation. Up to six seats each may be reserved.

Which seats will you book? Why?

Do you prefer to sit in the middle or at the end of the aisle?

Do you like the front rows or the back rows?

After each student has taken a turn, discuss the students' responses.

Were there any difficulties in booking?

Were you able to book the seats of your first choice?

What was the total number of seats that were booked?

How many seats were not booked?

My dream backyard

Materials

- 1 cm grid or square dot paper
- paper
- pencils/pens

Ask the students to draw a simple grid on a sheet of grid or dot paper and mark the coordinates. Use letters for the horizontal axis and numbers for the vertical axis.

Imagine that you are designing your dream backyard.

What do you think it would include?

Will your dream backyard have a paved BBQ area, a lawn, a vegetable garden, pot plants, a swimming pool, a pool deck, a spa, fences, a trampoline, a basketball ring, a clothesline or anything else?

Discuss the students' responses.

When the maps are complete, question the students concerning the coordinates of the various items on their backyard maps. Retain their work for the next activity.

Using scale

Materials

- copies of dream backyard plan from previous activity
- pencils/pens

Indicate to the students that they are now going to apply a scale of 1 cm to 1 m to their backyard design.

Can you choose a feature of your backyard and calculate its area?

Which item has the largest area? Which has the smallest area?

How many square metres does the lawn cover?

How many square metres does the swimming pool cover?

Discuss and display the students' efforts.

Encourage the students to devise some questions of their own and have their classmates answer them.

Street directory

Materials

- a computer
- screen image of local area
- multiple copies of map
- paper
- pencils/pens

Enter the link 'street-directory.com.au' online.

Enter the name of your school and discuss its position on the screen image. Discuss various landmarks, such as parks, shops, churches, football fields and netball courts.

Ask the students to choose two or more landmarks and mark them on their own maps.

Can you write some directions to get from one landmark to another?

Encourage the students to use directional language and compass directions to describe their route. Remind them of the convention that north is positioned at the top of the map.

Discuss the students' efforts. Retain their work for the next activity.

Extension

Ask the students to plan and record a route between multiple locations on the map.

Variation

Introduce the idea of scale found on the map and encourage the students to calculate the distances involved.

On the grid

Materials

- copies of local map from previous activity
- paper
- pencils/pens

Ask the students to draw a simple grid on their maps and mark the coordinates. Use letters for the horizontal axis and numbers for the vertical axis. Ensure that in this case the students mark the lines on the grid, rather than the spaces between the lines.

Using coordinates, can you identify the approximate position of: your school, a shop, a park, your house, a friend's house?

Select a student and ask them to choose a coordinate. Ask the class to identify the nearest street for that coordinate. Select other students and repeat the activity. Discuss the results and retain the students' maps for the next activity.

Grid routes

Materials

- copies of local map from previous activity
- paper
- pencils/pens

Imagine that you are a parcel delivery driver, making stops along the way.

Can you plan a route that has six or more stops?

At which locations will you stop?

Remember that your aim is to cover the route in the smallest possible distance.

Discuss the students' choice of routes. Choose individual students to explain their choices.

Can you now describe your route using the coordinates of the delivery stops?

Extension

Apply a scale to the students' maps and have them calculate the distances involved.

Year 6

Plot the shape

Materials

- a computer
- screen
- whiteboard
- markers
- 1 cm grid paper
- pencils/pens
- ruler

Display an 8 × 8 grid on the whiteboard or screen.

Label the points on the horizontal axis using the letters from A to H.

Label the points on the vertical axis using the numbers from 1 to 8.

Select individual students to help you plot coordinates as follows:

B2, B3, A4, B5, B6, C6, D7, E6, F6, F5, G4, F3, F2, E2, D1, C2, B2

Before joining the dots, ask the students to predict the shape that will be formed.

Encourage each student to draw a shape of their own and write the coordinates to describe it. Students can then swap their details with a classmate and ask them to create the shape.

Variation

Ask the students to sketch pictures on a sheet of grid paper and repeat the activity.

Quadrant plot

Materials

- a computer
- screen
- whiteboard
- markers
- 1 cm grid paper
- pencils/pens
- ruler

Display the quadrant diagram on the whiteboard or screen.

Explain to the students that coordinates may be plotted using positive and negative values along the x and y axes in all four quadrants. Point out that:

- if both values are positive, then the point lies in the first quadrant (1)
- if the x value is negative and the y value is positive, then the point lies in the second quadrant (2)
- if both values are negative, then the point lies in the third quadrant (3)
- if the x value is positive and the y value is negative, then the point lies in the fourth quadrant (4).

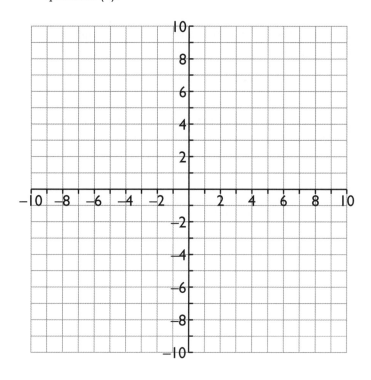

Label the quadrants on the diagram.

In which quadrant will you find (3, 2), (−3, 7), (−6, −8) and (5, −3)?

Discuss the students' responses.

Distribute the grid paper and ask the students to construct and label the quadrants.

Invite them to construct a polygon that covers all four quadrants and write the coordinates to define its shape. Ask the students to swap the coordinates with each other so that they can recreate each other's shapes.

Car plot

Materials

- a computer
- screen
- whiteboard
- markers
- 1 cm grid paper
- pencils/pens
- ruler

Distribute the grid paper and ask the students to construct and label the quadrants.

Invite them to sketch the outline of a car, covering all four quadrants and write the coordinates to define its shape. Ask the students to swap coordinates with each other so that they can recreate each other's car designs.

What on Earth?

Materials

- a world globe, or screen image

Display the globe or screen image to the students.

Explain to the students that latitude and longitude are the coordinate system of imaginary lines used to locate and describe the position of any place on earth.

Latitude is the distance north or south of the equator. Lines of latitude run parallel to the equator in the northern and southern hemispheres.

Meridians of longitude run at right angles to the lines of latitude, stretching from pole to pole. Longitude is measured eastwards and westwards from the Greenwich meridian.

Can you see the equator?

Can you see the northern and southern hemispheres?

What are the names of the areas at the top and bottom of the world globe?

Is Australia north or south of the equator?

Where does the Tropic of Capricorn cross Australia?

Variation

Ask the students to repeat the activity by referring to their atlases.

Extension

Investigate the position and nature of the International Date Line and how it relates to telling time on earth.

Investigating GPS

Materials

- a computer
- screen
- paper
- pencils/pens

Do you know what the initials GPS stand for?

Discuss the Global Positioning System and its use to describe the position of any place on earth.

Search the terms 'latitude' and 'longitude' online and select an appropriate application. Detail how latitude and longitude is recorded.

Ask the students to record their answers to the following questions.

What is the latitude and longitude of our school using GPS?

What is the latitude and longitude of our town or suburb?

What is the latitude and longitude of the nearest capital city?

Extension

Ask the students to investigate the coordinates of the place where they live and other locations of their choice.

Variation

Ask the students to investigate how the GPS works and which country provides the GPS for the world's use.

What in the world?

Materials

- a computer
- screen
- outline map of Australia
- pencils/pens

Search for 'Wikimapia' online.

Enter 'Australia' in the horizontal search bar.

Enter a city and press the plus or minus buttons on the vertical search bar to zoom in or out. Find the locations of Australia's capital cities on the map and discuss them with the students.

Ask the students to mark in the states and capital cities on an outline map of Australia. Have them record the GPS coordinates of each capital city.

Extension

Use Wikimapia to investigate other towns and cities in other countries and write their coordinates.

Where in the world?

Materials

- a computer
- screen

Go to Google Maps.

Select a location. Click on 'Map' and 'Satellite' in turn to demonstrate the different representations of the countryside to the students.

Can you see how maps can be made from photographs?

Do you know from where the photographs are taken?

Discuss how the photographic imagery is provided by a network of satellites covering the globe.

Click on the 'Directions' icon. Enter the name of the school in the starting point box.

Ask the students to suggest a possible end address such as a park, shops or a sports field.

Enter the chosen item in the destination box for the route to be displayed.

Change the route by adding a new destination.

Click on 'Add destination' to see the new route.

Extension

Have the students use the program to plan other journeys of their own. Print them out and discuss them with their classmates.

ANGLES

Year 5

Demonstrating angles

Materials

- geostrips or cardboard strips
- split pins

Join two geostrips together and display them to the students.

Turn the strips back and forth.

What have I made with these strips?

Call for answers. Discuss the concept of an angle as an amount of turn between two lines, and not the amount of space between the two lines.

OXFORD UNIVERSITY PRESS

Hidden angles

Materials

- a whiteboard
- markers

Hold the classroom door slightly open.

Look at the bottom edge of the door and the doorway?

Do you see a two-arm angle or a one-arm angle?

Explain to the students that sometimes, as in this example, the second arm of the angle cannot be seen.

Draw an inclined line on the whiteboard with a diagram of a car travelling down it.

Look at the car going down the slope.

Do you see a two-arm angle or a one-arm angle?

Further discuss this example. Ask the students to sketch and label the doorway and inclined car as one-arm angles formed.

Extension

Discuss examples of angles where no arms are evident, such as a ball bouncing off the cushions of a pool table or the original video game called Kong. Search online for examples of angles that are not clearly apparent.

Hidden shapes

Ask the students to close their eyes and visualise what is happening in the following problem.

A person is facing north. They take eight steps forwards and turn 90 degrees left. They take four steps forward and turn 90 degrees left again. They take another eight steps forward and turn 90 degrees left again. Finally, they take four steps forward and turn 90 degrees left.

In which direction are they now facing? What shape was traced by their path?

Call for answers and discuss the students' responses.

Human angles

Materials

- string or wool
- paper
- pencils/pens

Go outside the classroom and form groups of three students. Challenge them to create different types of angles using wool or string.

Can you make a right angle? An acute angle? An obtuse angle?

A straight angle? A full-turn angle?

Can you draw each of these angles?

Create the angle

Materials

- coloured paper circles
- scissors
- paper
- pencils/pens

Demonstrate to the students how to make an angle wheel. Take two different coloured paper circles and cut a slit to the centre of each paper circle. Line up the slits and then carefully fit the circles together to make a wheel.

Using your wheel, can you make an angle of 90 degrees?

Can you make an angle of 45 degrees?

What about 30 degrees? 60 degrees? 120 degrees? 180 degrees?

Display the angles made by the students. Label the angle measures.

Folding angles

Materials

- coloured paper circles

Challenge the students to create angles by paper folding.

Can you fold the paper to make an angle of 90 degrees? 60 degrees? 45 degrees? 30 degrees?

Discuss the students' efforts.

Construct the angle

Materials

- protractors
- square dot paper or A4 paper
- rulers
- pencils/pens

Tell the students that they are going to construct angles of different sizes.

Can you construct a right angle? An acute angle? An obtuse angle? A reflex angle?

Give directions and support where necessary. Ask the students to label the sizes of the angles involved.

Year 6

Hunt and measure

Materials

- geostrips or thin cardboard strips
- split pin paper fasteners
- protractors
- paper
- pencils/pens
- rulers

Ask the students to join two strips with paper fasteners to make angle testers.

Can you identify some angles outside of the classroom and measure their sizes?

Ensure that the students consider the natural as well as the built environment. Ask them to measure angles with the angle tester, transfer the angles using pencil and paper and measure them with the protractors.

On returning to the classroom, discuss the students' findings. Label and display their work.

Estimate and measure 1

Materials

- a whiteboard
- protractors
- paper
- rulers
- pencils/pens

Ask the students to copy angles like the above from the whiteboard.

Can you estimate the size of these angles in degrees?

Measure them using your protractor. Were your estimates accurate?

Angle game

Materials

- paper
- pencils/pens
- protractor
- 2 players

Divide the class into pairs and play the following game.

Ask the first student of the pair to draw a selection of 10 different angles on a sheet of paper. Both students then write their estimates next to each angle. They then use a protractor to measure the angles. The student who has made the largest number of best estimates is the winner.

Ask the second student to draw the angles and repeat the activity.

Angle on time

Materials

- toy clock or wall clock
- worksheets, each showing multiple blank analogue clock faces
- paper
- pencils/pens

Pose the following questions to the students.

Can you estimate the following angles made by the hands of the clock?

What is your estimate for 9 o'clock? Half past 6? 10 to 3? 20 past 9?

Allow time for the students to mark the times on their worksheets and determine their answers. Remind them to consider all types of angles made by the clock hands, including reflex angles.

Select different students to present their findings. Demonstrate the solutions to the class.

Extension

Ask class members to work in pairs. One student selects a time to the nearest 5-minute interval and challenges the other student to calculate the angles involved.

Trace and measure

Materials

- a whiteboard
- markers
- pattern blocks
- paper
- pencils/pens
- protractors

Ask the students to work in pairs and trace the different pattern blocks, using pencils and paper. Have them measure the size of the angles formed and calculate the angle sums.

Call for answers and record the results in a table on the whiteboard. Discuss the results.

Shape	Number of angles	Angle sizes	Angle sum
Triangle			
Square			
Trapezium			
Rhombus			
Hexagon			

Estimate and measure 2

Materials

- paper
- pencils/pens
- protractors

Ask the students to construct five different irregular polygons on their paper. Before measuring their angles, ask them to estimate the size of each of the angles and record them. Have them calculate the angle sum of each polygon.

What did you find out?

Were your estimates accurate?

Discuss the results.

OXFORD UNIVERSITY PRESS

Opposite angles

Materials

- whiteboard
- markers
- pencils/pens
- paper
- rulers
- protractors

Draw a pair of intersecting lines on the whiteboard and ask the students to study them.

What do you notice about the angles made by these intersecting lines?

Allow sufficient time for the students to talk about their ideas with each other.

Call for answers and discuss the students' ideas.

Ask the students to construct their own sets of intersecting lines and measure the angles involved with a protractor.

What did you find out?

How many pairs of equal angles did you find?

What is the sum of all the angles?

Discuss how vertically opposite angles are always equal. In this case, vertical refers to the vertex or crossing point of the lines and not an up and down direction. The angle sum is 360 degrees.

Extension

Investigate opposite angle problems in which the size of one of the four angles is given and the other three angles must be calculated.

Imagine this

Encourage the students to practise their visualisation skills by answering the following questions.

Close your eyes and imagine your answers to these questions:

- *I am an equilateral triangle. One of my angles is 60 degrees. What is the size of the other two angles?*
- *I am an isosceles triangle with two angles of 70 degrees each. What is my other angle?*
- *I am a scalene triangle. One of my angles is 30 degrees and another is 45 degrees. What is the size of my other angle?*
- *I am a regular hexagon. What is the size of each of my angles?*
- *I am a quadrilateral with angles of 140 degrees, 60 degrees and 40 degrees. What is the size of my missing angle?*
- *I am a parallelogram. Two of my angles are 50 degrees and 130 degrees. What are the sizes of my other two angles?*

Extension

Encourage the students to devise similar questions of their own.